in a week

Catherine Brown
and Lee Cope,
Abbey College, Birmingham
Series Editor: Kevin Byrne

Mathematics

Where to find the information you need

SUCCESS OR YOUR MONEY BACK

Letts' market leading series AS in a Week gives you everything you need for exam success. We're so confident that they're the best revision books you can buy that if you don't make the grade we will give you your money back!

HERE'S HOW IT WORKS

Register the Letts AS in a Week guide you buy by writing to us within 28 days of purchase with the following information:

- Name
- Address
- Postcode
- Subject of AS in a Week book bought

Please include your till receipt

To make a **claim**, compare your results to the grades below. If any of your grades qualify for a refund, make a claim by writing to us within 28 days of getting your results, enclosing a copy of your original exam slip. If you do not register, you won't be able to make a claim after you receive your results.

CLAIM IF...

You are an AS (Advanced Subsidiary) student and do not get grade E or above.

You are a Scottish Higher level student and do not get a grade C or above.

This offer is not open to Scottish students taking SCE Higher Grade, or Intermediate qualifications.

Registration and claim address:
Letts Success or Your Money Back Offer, Letts Educational,
414 Chiswick High Road, London W4 5TF

TERMS AND CONDITIONS

1. Applies to the Letts AS in a Week series only
2. Registration of purchases must be received by Letts Educational within 28 days of the purchase date
3. Registration must be accompanied by a valid till receipt
4. All money back claims must be received by Letts Educational within 28 days of receiving exam results
5. All claims must be accompanied by a letter stating the claim and a copy of the relevant exam results slip
6. Claims will be invalid if they do not match with the original registered subjects
7. Letts Educational reserves the right to seek confirmation of the level of entry of the claimant
8. Responsibility cannot be accepted for lost, delayed or damaged applications, or applications received outside of the stated registration/claim timescales
9. Proof of posting will not be accepted as proof of delivery
10. Offer only available to AS students studying within the UK
11. SUCCESS OR YOUR MONEY BACK is promoted by Letts Educational, 414 Chiswick High Road, London W4 5TF
12. Registration indicates a complete acceptance of these rules
13. Illegible entries will be disqualified
14. In all matters, the decision of Letts Educational will be final and no correspondence will be entered into

Letts Educational
Chiswick Centre
414 Chiswick High Road
London W4 5TF
Tel: 020 8996 3333
Fax: 020 8743 8390
e-mail: mail@lettsed.co.uk
website: www.letts-education.com

First published 2000
New edition 2004
10 9 8 7 6 5 4 3 2

British Library Cataloguing in Publication Data
A CIP record for this book is available from the British Library.

ISBN 1 84315 357 2

Cover design by Purple, London

Prepared by *specialist* publishing services, Milton Keynes
Design and project management by Starfish DEPM, London.

Printed in Italy

Letts Educational Limited is a division of Granada Learning Limited, part of Granada plc.

Algebra

25 mins
Time Yourself

How much do you know?

DAY 1

1 a) Solve the equation $3x^2 - 4x - 9 = 0$
b) Without solving the equation, state the number of solutions of the equation $x^2 - 6x + 10 = 0$, giving a reason for your answer.
c) Complete the square for $3x^2 - 12x + 2$

2 Solve the following pair of simultaneous equations:
$x + 4y = 6$
$x^2 + 2xy - 3y^2 = 5$

3 a) i) Express the following as powers of 3: 9^x, 27^{1-4x}
ii) Hence solve the equation $3^{2-3x}\,9^x = 27^{1-4x}$
b) Solve the equation $6^{2x} - 42(6^x) + 216 = 0$

4 Express the following in terms of the simplest possible surds:
a) $\sqrt{343}$ b) $\sqrt{768}$ c) $\dfrac{\sqrt{3}}{2\sqrt{6}}$ d) $\dfrac{3}{1-\sqrt{7}}$

5 Solve the following inequalities: a) $3x - 5 \geq 4x - 8$ b) $4x^2 - 8x + 3 > 0$

6 Show that $(2x-1)$ is a factor of $8x^3 - 12x^2 + 6x - 1$ and hence factorise this expression fully.

Answers

1a) 2.52, −1.19 b) $b^2 - 4ac = (-6)^2 - 4 \times 1 \times 10 = -4 < 0$, so no solutions c) $3(x-2)^2 - 10$ **2** $x = 2, y = 1$ and $x = -18.8, y = 6.2$ **3.** a) i) 3^{2x}, 3^{3-12x} ii) $x = \frac{1}{11}$ b) $x = 1, 2$
4 a) $7\sqrt{7}$ b) $16\sqrt{3}$ c) $\dfrac{\sqrt{2}}{4}$ d) $\dfrac{-(1+\sqrt{7})}{2}$
5 a) $x \leq 3$ b) $x > 1.5$ or $x < 0.5$ **6** Sub. in $x = \frac{1}{2}$: $8(\frac{1}{2})^3 - 12(\frac{1}{2})^2 + 6(\frac{1}{2}) - 1 = 1 - 3 + 3 - 1 = 0$, so $(2x-1)$ a factor. $8x^3 - 12x^2 + 6x - 1 \equiv (2x-1)(4x^2 - 4x + 1) \equiv (2x-1)^3$

If you got them all right, skip to page 10

2 3 4 5 6 7

Algebra

Learn the key facts

1 Quadratics

You need to be able to do the following with quadratics:

- Solve quadratic equations (see *GCSE Maths to A* In a Week*)
- Draw graphs of quadratics
- Decide whether quadratic equations have no roots, equal roots or two roots
- Complete the square.

To decide how many roots quadratic equations have, we look at the discriminant.

Discriminant $= b^2 - 4ac$

This is what is under the square root sign in the quadratic formula.

- If the discriminant is positive, there are two different roots
- If the discriminant is zero, there is one root (which is a repeated root)
- If the discriminant is negative, there are no roots.

This is because we can't take the √ of a negative number

Completing the square means writing a quadratic expression in the form $A(x + B)^2 + C$, where A, B and C are numbers. There are several ways to do it – here's one:

Example 1

Complete the square for $2x^2 + 12x + 7$

Solution

Set $2x^2 + 12x + 7 \equiv A(x + B)^2 + C$

Now expand the brackets: $2x^2 + 12x + 7 \equiv Ax^2 + 2ABx + AB^2 + C$

We now look at the x^2 terms, the x terms and the constant terms:

x^2 terms:	$2 = A$
x terms:	$12 = 2AB$, so $B = 3$
Constant:	$7 = AB^2 + C \Rightarrow C = -11$
Answer:	$2(x + 3)^2 - 11$

Anything with no x in it is a constant term

2 Simultaneous Equations

From GCSE, you should know how to solve standard simultanous equations. For AS, you also need to be able to solve them when one is linear (i.e. 'normal') and one is quadratic.

To do this, you rearrange the linear one to make y (or x) the subject, substitute into the harder one, multiply out and solve the quadratic equation you get.

Example 2

Solve the simultaneous equations: $y - 2x = 1$ **(1)** $x^2 + 2xy + y^2 = 16$ **(2)**

Solution

(1) $\Rightarrow y = 2x + 1$

Substitute into **(2)**: $x^2 + 2x(2x + 1) + (2x + 1)^2 = 16$
$\Rightarrow x^2 + 4x^2 + 2x + 4x^2 + 4x + 1 = 16$

Simplifying: $9x^2 + 6x - 15 = 0$

Solving: $x = \frac{-6 \pm \sqrt{6^2 - 4 \times 9 \times -15}}{2 \times 9}$ so $x = 1$ or $-\frac{5}{3}$. If $x = 1$, $y = 3$. If $x = -\frac{5}{3}$, $y = -\frac{7}{3}$

Don't forget to substitute back to find y!

3 Powers

You need to learn the laws of powers, which are:

You MUST have the same base for each power

$$y^m \times y^n = y^{m+n} \qquad \frac{y^m}{y^n} = y^{m-n} \qquad (y^m)^n = y^{mn}$$

In other words: if you are multiplying, add the powers
if you are dividing, subtract them
if you have a power of a power, you multiply.

In addition to the laws, you need to know that:

Anything to the power 0 is 1.

Anything to the power 1 is the number itself – e.g. $5^1 = 5$

LEARN!

A negative power means 'one over' – so $x^{-3} = \frac{1}{x^3}$

A fraction power means roots – so $x^{\frac{1}{2}} = \sqrt{x}$; $x^{\frac{1}{3}} = \sqrt[3]{x}$ $x^{\frac{3}{2}} = \sqrt{x^3}$ or $(\sqrt{x})^3$

You can only manipulate powers in certain ways.

You can take out brackets when only $\times$ or $\div$ are involved, but not with + or –.

For example: $(2x)^3 = 2^3x^3$; $\left(\frac{5}{3}\right)^4 = \frac{5^4}{3^4}$ but $(x + a)^{0.5}$ is definitely NOT $x^{0.5} + a^{0.5}$.

Examples 3 and 4 illustrate two ways of using these facts about powers; they are also essential in many other areas.

DAY 1 2 3 4 5 6 7

Example 3

a) Express i) 2^{4x-6} ii) 8^{2x-4} in the form 4^y, where y is to be determined in terms of x.

b) Hence solve the equation $2^{4x-6} = \dfrac{4^{2-x}}{8^{2x-4}}$

We must have both sides in the form $2^{\text{something}}$ before doing this

Solution

a) i) Set $2^{4x-6} = 4^y$ (since we are told to).

To solve this, we need both sides as powers of the same number.

So we use $4 = 2^2$, so $2^{4x-6} = 4^y$ becomes $2^{4x-6} = (2^2)^y = 2^{2y}$.

Since both sides are powers of the same number, equate powers:

$4x - 6 = 2y \Rightarrow y = 2x - 3$, so $2^{4x-6} = 4^{2x-3}$

ii) $8^{2x-4} = 4^y \Rightarrow (2^3)^{2x-4} = (2^2)^y \Rightarrow 2^{6x-12} = 2^{2y} \Rightarrow 6x - 12 = 2y \Rightarrow y = 3x - 6$

b) We must use what we've done: $2^{4x-6} = \dfrac{4^{2-x}}{8^{2x-4}} \Rightarrow 4^{2x-3} = \dfrac{4^{2-x}}{4^{3x-6}}$

We CAN'T cancel the fours yet!

Since we are dividing, we subtract powers:
$4^{2x-3} = 4^{2-x-(3x-6)} \Rightarrow 4^{2x-3} = 4^{-4x+8} \Rightarrow 2x-3 = -4x + 8 \Rightarrow x = \frac{11}{6}$

Example 4

a) Given that $y = 5^x$, show that i) $125^x = y^3$ ii) $25^x = y^2$ iii) $5^{x+1} = 5y$

b) Hence solve the equation $125^x - 6(25^x) + 5^{x+1} = 0$

Solution

a) i) $125^x = (5^3)^x = 5^{3x}$ $\quad y^3 = (5^x)^3 = 5^{3x}$ So $125^x = y^3$

ii) $25^x = (5^2)^x = 5^{2x}$ $\quad y^2 = (5^x)^2 = 5^{2x}$ So $25^x = y^2$

iii) $5^{x+1} = 5^x \times 5^1$ $\quad 5y = 5 \times 5^x = 5^x \times 5^1$ So $5^{x+1} = 5y$

b) Using what we've done: $125^x - 6(25^x) + 5^{x+1} = 0 \Rightarrow y^3 - 6y^2 + 5y = 0$

Factorising: $y(y^2 - 6y + 5) = 0 \Rightarrow y(y - 5)(y - 1) = 0$

So $y = 0$, or $y = 5$, or $y = 1$. But $y = 5^x \Rightarrow 5^x = 0$, or $5^x = 5$, or $5^x = 1$

$\Rightarrow 5^x = 0$ gives no solution, $5^x = 5$ gives $x = 1$ and $5^x = 1$ gives $x = 0$

4 Surds

A surd is something with a square root in it – like $\sqrt{2}$ or $\sqrt{5}$. You need to be able to manipulate surds in the following ways.

Simplifying: this is when you have the square root of a number and you have to express it in terms of the simplest possible surds – which will usually be things like $\sqrt{2}$, $\sqrt{3}$ or $\sqrt{5}$. Example 5 illustrates this.

Example 5

Express in terms of the simplest possible surds: a) $\sqrt{8}$ b) $\sqrt{243}$

Solution

a) We look for a square number that goes into 8; 4 does.

So write: $\sqrt{8} = \sqrt{(4 \times 2)} = \sqrt{4} \times \sqrt{2} = 2\sqrt{2}$

b) Finding a square number again: 9 goes into 243.

So write: $\sqrt{243} = \sqrt{(9 \times 27)} = \sqrt{9} \times \sqrt{27} = 3\sqrt{27}$

But we haven't finished, because there's a number that goes into 27; 9 again:

So $\sqrt{243} = 3\sqrt{27} = 3\sqrt{(9 \times 3)} = 3\sqrt{9} \times \sqrt{3} = 3 \times 3\sqrt{3} = 9\sqrt{3}$

Rationalising the denominator: in other words getting surds off the bottom of a fraction. This is done by multiplying the top and bottom of the fraction by the same thing. There are two cases – according to what the denominator is like. These are illustrated in Example 6.

Example 6

Rationalise: a) $\dfrac{2}{3\sqrt{5}}$ b) $\dfrac{4-2\sqrt{6}}{2+\sqrt{3}}$

Solution

a) Multiply top and bottom by $\sqrt{5}$: $\dfrac{2}{3\sqrt{5}} \times \dfrac{\sqrt{5}}{\sqrt{5}} = \dfrac{2\sqrt{5}}{3\sqrt{5}\sqrt{5}} = \dfrac{2\sqrt{5}}{3 \times 5} = \dfrac{2\sqrt{5}}{15}$

b) Multiply top and bottom by $2 - \sqrt{3}$:

$$\frac{(4-2\sqrt{6})}{(2+\sqrt{3})} \times \frac{(2-\sqrt{3})}{(2-\sqrt{3})} = \frac{(4-2\sqrt{6})(2-\sqrt{3})}{(2+\sqrt{3})(2-\sqrt{3})} = \frac{8-4\sqrt{3}-4\sqrt{6}+2\sqrt{6}\sqrt{3}}{4-2\sqrt{3}+2\sqrt{3}-\sqrt{3}\sqrt{3}} = \frac{8-4\sqrt{3}-4\sqrt{6}+2\sqrt{18}}{4-3}$$

$= 8 - 4\sqrt{3} - 4\sqrt{6} + 2\sqrt{9}\sqrt{2} = 8 - 4\sqrt{3} - 4\sqrt{6} + 6\sqrt{2}$

5 Inequalities

When dealing with inequalities, you must remember the following rules:

- You can add or subtract anything whatsoever to either side
- You can multiply or divide by positive numbers
- You can multiply or divide by negative numbers, provided you change the direction of the sign
- You cannot multiply or divide by unknowns – like x
- You cannot square, square root, turn them upside down – or anything else!

Be careful with $<$ or $\le$ – getting it wrong will cost marks!

Linear Inequalities

These are easy! You treat them just like equations (but remember about multiplying/dividing by negatives).

Example 7

Solve the following inequality: $2x - 5 \le 3x - 4$

Solution

Take x terms to one side, numbers to the other:

$2x - 5 \le 3x - 4 \Rightarrow 2x - 3x \le -4 + 5 \Rightarrow -x \le 1$

Get rid of the minus sign:

$-x \le 1 \Rightarrow x \ge -1$

Quadratic Inequalities

To deal with these, you first rearrange it so you have 0 on one side, then find the roots and then draw a number line; the method is illustrated in Example 8.

Example 8

Solve the following inequality: $x^2 - 3x - 4 > 0$

Solution

STEP 1: Find roots of the equation:

$x^2 - 3x - 4 = 0 \Rightarrow (x - 4)(x + 1) = 0 \Rightarrow x = 4, -1$

STEP 2: Mark them on a number line

−1 4

STEP 3: Choose values of x each side of the roots and between them; put them into the quadratic

$x = -2$: $(-2)^2 - 3(-2) - 4 = 6$ +ve

$x = 0$: $(0)^2 - 3(0) - 4 = -4$ −ve

$x = 5$: $(5)^2 - 3(5) - 4 = 6$ +ve

STEP 4: Mark the appropriate regions on the number line +ve or −ve

+ve −1 −ve 4 +ve

STEP 5: Write down the answerWe want it >0, so we want $x < -1$ or $x > 4$

6 Factor Theorem

The factor theorem says: $(x - a)$ is a factor of $f(x) \Leftrightarrow f(a) = 0$

We use this to factorise cubic equations.

- If you are asked to show something is a factor, you just substitute in the appropriate number and show the answer is equal to zero.
- If you have to find a factor, you guess ± numbers that go into the constant term – and then put them in to see if you get zero.
- Once you have found one factor, don't carry on guessing! You then factorise.

Whenever you factorise a cubic, you MUST use this method!

The method is demonstrated in Example 9.

Example 9

Factorise fully: $x^3 + 3x^2 - 13x - 15$

Solution

Find one factor by guessing. We guess ±1, ±3, ±5, ±15

$x = 1$: $(1)^3 + 3(1)^2 - 13(1) - 15 = -24$ so $(x - 1)$ not a factor

$x = -1$: $(-1)^3 + 3(-1)^2 - 13(-1) - 15 = 0$ so $(x + 1)$ a factor

Now set $x^3 + 3x^2 - 13x - 15 \equiv (x + 1)$(quadratic).

We need to find the terms of the quadratic.

First term must be x^2 (since $x \times x^2$ is needed to get x^3).

Last term must be -15 (since 1×-15 is needed to get -15).

Now we must find the middle term – which we will call Ax:

So $x^3 + 3x^2 - 13x - 15 \equiv (x + 1)(x^2 + Ax - 15)$.

We now look at the x^2 terms on both sides:

On the left, we have $3x^2$. On the right, we get x^2 terms from $1 \times x^2$ and $x \times Ax$

So $3x^2 \equiv x^2 + Ax^2 \Rightarrow 3 = 1 + A$ So $A = 2$

So $x^3 + 3x^2 - 13x - 15 \equiv (x + 1)(x^2 + 2x - 15) \equiv (x + 1)(x + 5)(x - 3)$

Algebra

Have you improved?

1 a) Find the solutions to the quadratic equation $x^2 - 6x + 2 = 0$, giving your answers in terms of the simplest possible surds.
b) Hence obtain the exact solution to the inequality $x^2 - 6x + 2 \leq 0$

2 A gardener is planning a rectangular lawn. He requires its width to be 4 m less than its length, its perimeter to be at most 36 m and its area to be at least 60 m^2. By forming and solving suitable inequalities, find the range of acceptable values for the length of the lawn.

3 $f(x) \equiv -2x^2 - 20x + 15$.
a) Express $f(x)$ in the form $A(x + B)^2 + C$, where A, B and C are constants to be determined.
b) Hence:
i) Obtain the solution to the equation $f(x) + 31 = 0$, giving your answer in terms of the simplest possible surds.
ii) State the maximum value of $f(x)$.

4 By using the substitution $y = 2^x$, obtain the solution to the equation $2^{x+1} + 2^{-x} = 3$

5 a) Show that the cubic equation $x^3 - x^2 - x - 2 = 0$ has only one real root.
b) Hence solve $2^{3x} - 2^{2x} - 2^x - 2 = 0$

1a) Use the formula Simplify the square root 1b Even though it's surds, use the normal method

2) Call the length x – what will the width be? Work out the perimeter and area in terms of x, and use these to get the inequalities You need both inequalities to be true at once!

3a) Look back at the method for completing the square! 3bi) Put what you got in a) into this equation to replace f(x) Rearrange to make the $(x + B)^2$ the subject, then square root. Don't forget ± 3bii) The squared part can only be 0 or positive

4) $2^{x+1} = 2x \times 2^1$ $2^{-x} = 1/2^x$ After substituting in, multiply by y

5a) Try to factorise it as normal. Look at how many roots the quadratic has 5b) Put $y = 2^x$

Coordinate Geometry

30 mins
Time Yourself

How much do you know?

1 A straight line *l* passes through A(2, −8), B(−1, 7) and C(k, 4.5) meeting the coordinate axes at D and E.

a) Calculate the equation of the line *l* in the form $y = mx + c$.

b) Deduce the value of k.

c) Find the length DE to 3 significant figures.

2 The points P, Q and R have coordinates (1, 0), (3, 4) and (−3, 2) respectively.

a) Show that the triangle PQR is isosceles.

Given that M is the midpoint between Q and R then:

b) Calculate the distance PM,

c) Hence calculate the area of the triangle PQR,

d) Deduce $\tan P\hat{Q}R$.

3 Calculate the coordinates where the line $y = 4x + 9$ intersects the curve $y = 3x^2 + x - 27$.

4 The points A, B and C have coordinates (5, 6), (4, −2) and (6, 0) respectively. Expressing your answers in the form $ax + by + c = 0$, where a is a positive integer, find the equation of the:

a) straight line n perpendicular to AB and which passes through the midpoint of BC

b) straight line *l* parallel to n, which passes through the point C.

Answers

1 a) $y = -5x + 2$ b) -0.5 c) 2.04

2 a) $PQ = \sqrt{20}$, $PR = \sqrt{20}$, $QR = \sqrt{40} \Rightarrow PQ = PR \neq QR \Rightarrow \triangle PQR$ is isosceles
b) $\sqrt{10}$ c) 10 (units)^2 d) 1

3 (4, 25) and (−3, −3)

4 a) $x + 8y + 3 = 0$ b) $x + 8y - 6 = 0$

If you got them all right, skip to page 15

Spend no more than 60 mins on this topic

Coordinate Geometry

DAY 1

Learn the key facts

1 Equation of a Line

There are three ways you can write an equation of a line. They are $y = mx + c$, $y - y_1 = m(x - x_1)$, or $ax + by + c = 0$. The best and easiest formula to use is $y - y_1 = m(x - x_1)$, where m is the gradient of the line, and (x_1, y_1) is a point on the line.

THINK: You need a GRADIENT and a POINT to find a LINE.

Example 1

Express the equation of the line joining the two points (4, 3) and (−4, 8) in the form $ax + by + c = 0$, where a is a positive integer.

Solution

Gradient and point ⇒ Line.

We need the gradient, $m = \dfrac{\text{change in } y}{\text{change in } x} = \dfrac{y_1 - y_2}{x_1 - x_2}$, for points (x_1, y_1) and (x_2, y_2).

So $m = \dfrac{3-8}{4--4} = \dfrac{-5}{8}$, and the point = (4, 3) say.

Hence the formula ⇒ $y - 3 = \dfrac{-5}{8}(x-4)$ $\quad (\times 8)$

Hence $8y - 24 = -5x + 20 \Rightarrow 8y - 24 + 5x - 20 = 0$
giving $5x + 8y - 44 = 0$, as required.

2 Distance Between Two Points

The distance between two points is represented by:

$\sqrt{(x_1 - x_2)^2 + (y_1 - y_2)^2}$ **in 2-D** for points (x_1, y_1) & (x_2, y_2).

Example 2

A triangle ABC has coordinates A(−4, −3), B(−7, 1) and C(2, 1.5).

a) Prove that the angle $B\hat{A}C$ is 90°.

b) Find sin $A\hat{C}B$, giving your answer to three significant figures.

c) Find the length, to three significant figures of the perpendicular from A to BC.

Solution

a) Distance AB = $\sqrt{(-4-(-7))^2+(-3-1)^2} = \sqrt{9+16} = \sqrt{25} = 5$

distance BC = $\sqrt{(-7-2)^2+(1-1.5)^2} = \sqrt{81+0.25} = \sqrt{81.25}$

distance AC = $\sqrt{(-4-2)^2+(-3-1.5)^2} = \sqrt{36+20.25} = \sqrt{56.25}$

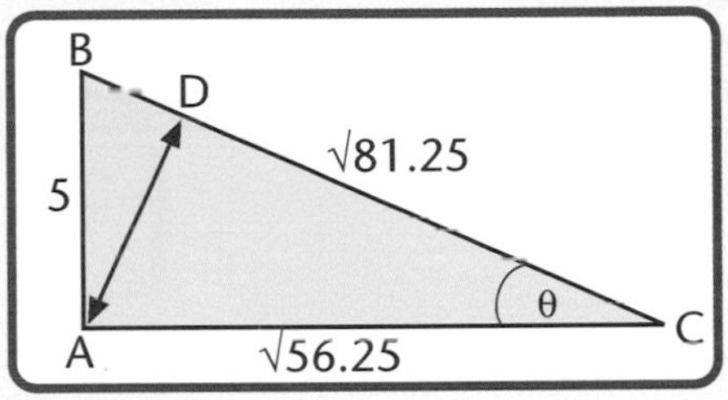

From the diagram, $AB^2 + AC^2 = BC^2$, by Pythagoras' Theorem for right-angled triangles.

LHS = $(5)^2 + (\sqrt{56.25})^2 = 25 + 56.25 = 81.25 = (\sqrt{81.25})^2 = BC^2 = RHS$.

$\therefore$ Angle $B\hat{A}C = 90^\circ$, by Pythagoras' Theorem.

b) Angle required is marked as θ, so

$$\sin\theta = \frac{OPP}{HYP} = \frac{5}{\sqrt{81.25}} = 0.5547... = 0.555(3\,sf)$$

c) Let D be the foot of the perpendicular from A to BC, as marked in the diagram.

So $A\hat{D}C = 90^\circ$, and using trigonometry, we have, $\sin\theta = \frac{OPP}{HYP} = \frac{AD}{AC} = \frac{AD}{\sqrt{56.25}}$

So distance required = AD = $\sqrt{56.25}\sin\theta = \sqrt{56.25} \times 0.5547... = 4.16(3\,sf)$

The MIDPOINT, M, of the line joining the two points $A(x_1, y_1)$ and $B(x_2, y_2)$ is found by averaging their two coordinates

$$M\left(\frac{x_1+x_2}{2}, \frac{y_1+y_2}{2}\right)$$

3 Intersecting Lines and Curves

When lines intersect curves, we solve the two equations simultaneously.

Example 3

The line $y = 2x + 5$ intersects the curve $y = 2x^2 - 5x + 1$, at the points X and Y. Calculate the coordinates of the midpoint M, of the line segment XY.

Solution

Finding the coordinates X and Y ⇒ solve equations simultaneously.

Hence $2x + 5 = 2x^2 - 5x + 1$, giving $2x^2 - 7x - 4 = 0$.

Factorising gives: $(2x + 1)(x - 4) = 0 \Rightarrow x = -\frac{1}{2}$ or $x = 4$.

When $x = -\frac{1}{2}$, $y = 2(-\frac{1}{2}) + 5 = 4$.

When $x = 4$, $y = 2(4) + 5 = 13$.

So $X(-\frac{1}{2}, 4)$ and $Y(4, 13)$.

Finally, we find the midpoint: $M = \left(\dfrac{-0.5+4}{2}, \dfrac{4+13}{2}\right) \Rightarrow M(1.75, 8.5)$

4 Parallel and Perpendicular Lines

Parallel lines have the same gradient. Ie. $m_1 = m_2$

Gradients of perpendicular lines multiply to make −1. Ie. $m_1 m_2 = -1$

E.g. The line l: $y = -4x - 3$, has gradient $m = -4$

Hence a line parallel to l has gradient -4

A line perpendicular to l has gradient $\dfrac{-1}{-4} = \dfrac{1}{4}$

Example 4

The line l has equation $3x + 4y - 5 = 0$. The line m passes through the point A(−5, 5) and is perpendicular to l. The line n is parallel to m and passes through the midpoint of A and the point B(−7, −1).
Give the equations of (a) the line m, and (b) the line n, in the form $y = mx + c$.

Solution

a) We find the gradient of l by writing it in the form $y = mx + c$.

Hence $4y = -3x + 5 \Rightarrow y = \dfrac{-3}{4}x + \dfrac{5}{4}$

So the gradient of l is $-\dfrac{3}{4} \Rightarrow$ gradient of $m = \dfrac{-1}{-3/4} = \dfrac{4}{3}$

because l is perpendicular to m.

Gradient and point ⇒ $y - 5 = \dfrac{4}{3}(x - -5) \Rightarrow y = \dfrac{4}{3}x + \dfrac{35}{3}$. (Line m)

b) n is parallel to m ⇒ gradient $n = \dfrac{4}{3}$ and point $= \left(\dfrac{-5+-7}{2}, \dfrac{5+-1}{2}\right) = (-6, 2)$

Gradient and point ⇒ $y - 2 = \dfrac{4}{3}(x - -6) \Rightarrow y - 2 = \dfrac{4}{3}x + 8 \Rightarrow y = \dfrac{4}{3}x + 10$. (Line n)

Have you Improved?

1 Statto, a couch potato football fan, was asked to design the perfect chair for watching football on his television. Statto got some graph paper out and drew a side view of his fantasy chair. He used the scale of one unit in the *x*- and *y*-axis to represent 20 cm.

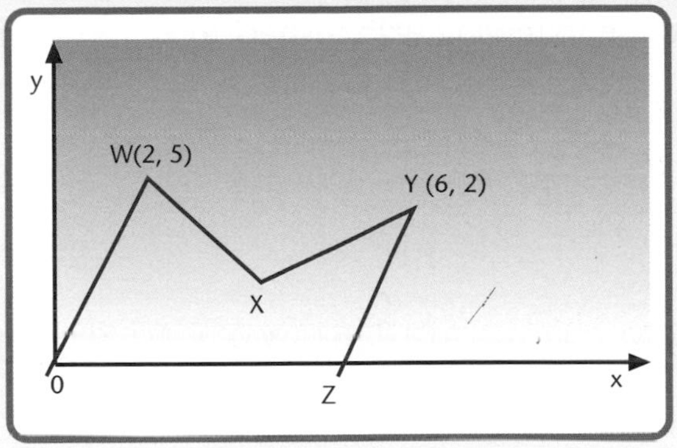

To begin, Statto drew a line from the origin to the point W, with coordinates (2, 5).

a) Write down the equation of the line OW.

Next Statto drew a line WX, with gradient −2, as shown in the diagram, which is perpendicular to the line XY, with Y having coordinates (6, 2).

b) Calculate the equation of the line XY.

c) Hence calculate the coordinates of X.

Finally the line YZ was drawn which is parallel to the line OW.

d) Calculate the length the side view occupies on the floor. (Take the *x*-axis as the floor.)

2 A straight line *l* passes through the points A(−3, 1) and B(*t*, −*t*) (where $t > 0$); meeting the *x*- and *y*-axis at the points P and Q respectively. The distance between the points A and B is 10 units.

Find:

a) the value of *t*

b) the equation of the line *l*

c) the area of the triangle OPQ, where O is the origin.

1a) Gradient and Point makes line

1b) Work out WX first, followed by WY using $m_1.m_2 = -1$

1c) Equate the lines WX & WY

1d) Work out equation YZ. Then set $y = 0$ to find where it cuts the x-axis

2a) Use distance formula AB^2. Form a quadratic. Solve!

2b) Gradient and Point

2c) Area = $\frac{1}{2}bh$

60 mins
Time Yourself

Differentiation

How much do you know?

1 Differentiate the following with respect to x:

a) $y = 5$ b) $y = 7x - 8$ c) $y = 8x^3$

d) $y = 3 - 2x - x^2$ e) $y = 4x - 3x^2 - 2x^3$ f) $y = 8x^3 - 3x - 2$

g) $y = (x + 7)(x + 4)$ h) $y = (2x + 3)(x - 2)$ i) $y = x(2x + 3)(x - 1)$

j) $y = 6\sqrt{x}$ k) $y = \sqrt[3]{x}$ l) $y = \sqrt{x}(4 - 3x)$

m) $y = \dfrac{2x^3 + x^2 - 3}{x}$ n) $y = \dfrac{2x^2 + 3x + 4}{x^{3/2}}$ o) $y = \dfrac{4 + 5x^2}{2x^2}$

2 Find the equations of a) the tangent and b) the normal, to the curve with equation $y = 2x^3 - 3x + 5$ at the point Q, where $x = -1$.

3 Find the complete set of values of x for which the following curves are increasing:

a) $y = (4 - x)(3 + x)$ b) $y = x^3 - 4x^2 + 4x - 5$ c) $y = x^2 + \dfrac{16}{x}$

4 Find the coordinates and determine the nature of the stationary points for the graph of $y = 2 + 15x - 6x^2 - x^3$.

Answers

1 a) 0 b) 7 c) $24x^2$ d) $-2 - 2x$ e) $4 - 6x - 6x^2$ f) $24x^2 - 3$ g) $2x + 11$
h) $4x - 1$ i) $6x^2 + 2x - 3$ j) $3x^{-1/2}$ k) $\frac{1}{3}x^{-2/3}$ l) $2x^{-1/2} - \frac{9}{2}x^{1/2}$
m) $4x + 1 + 3x^{-2}$ n) $x^{-1/2} - \frac{3}{2}x^{-3/2} - 6x^{-5/2}$ o) $-4x^{-3}$
2a) $y = 3x + 9$ b) $y = -\frac{1}{3}x + \frac{17}{3}$ **3a)** $x < \frac{1}{2}$ b) $x < \frac{2}{3}$ or $x > 2$ c) $x > 2$
4 Minimum SP at $(-5, -98)$, Maximum SP at $(1, 10)$

If you got them all right, skip to page 24

DAY 1 2 3 4 5 6 7

Spend no more than 120 mins on this topic

Learn the key facts

The notation for the first derivative used by exam boards is $\frac{dy}{dx}$ or $f'(x)$.

The notation for the second derivative is $\frac{d^2y}{dx^2}$ or $f''(x)$.

$\frac{d^2y}{dx^2}$ or $f''(x)$ means you differentiate a function $f(x)$ twice.

1 Basic Differentiation

In AS Level mathematics you are expected to differentiate the following type of equation:

When $y = x^n$ then $\frac{dy}{dx} = nx^{n-1}$ where n is a real number.

Bring the power down, and reduce the power by one

Example 1

Differentiate the following with respect to x:

a) $y = x^7$ b) $y = 5x^4$ c) $y = 4x^2 - 3x + 2$

d) $y = (x + 7)(x - 4)$ e) $y = (2x - 3)(x + 2)$ f) $y = \frac{4}{3x^2}$

g) $y = \frac{x^3 + 3x^4}{x^2}$ h) $y = \frac{3x^7 - x}{x^5}$ i) $y = \frac{4x^2 + 2x - 5}{\sqrt{x}}$

Solution

The first three are fairly straightforward:

a) $\frac{dy}{dx} = 7x^6$ (b) $\frac{dy}{dx} = 5 \times 4x^3 = 20x^3$ c) $\frac{dy}{dx} = 8x - 3$

You must manipulate the next six before you can differentiate:

d) $y = x^2 - 4x + 7x - 28 = x^2 + 3x - 28 \Rightarrow \frac{dy}{dx} = 2x + 3$

e) $y = 2x^2 + 4x - 3x - 6 = 2x^2 + x - 6 \Rightarrow \frac{dy}{dx} = 4x + 1$

DAY 1 2 3 4 5 6 7

1
DAY
2
3
4
5
6
7

f) $y = \frac{4}{3}x^{-2} \Rightarrow \frac{dy}{dx} = \frac{-8}{3}x^{-3} = \frac{-8}{3x^3}$

g) $y = \frac{x^3}{x^2} + \frac{3x^4}{x^2} = x + 3x^2 \Rightarrow \frac{dy}{dx} = 1 + 6x$

h) $y = \frac{3x^7}{x^5} - \frac{x}{x^5} = 3x^2 - x^{-4} \Rightarrow \frac{dy}{dx} = 6x + 4x^{-5}$

i) $y = \frac{4x^2}{x^{1/2}} + \frac{2x}{x^{1/2}} - \frac{5}{x^{1/2}} = 4x^{3/2} + 2x^{1/2} - 5x^{-1/2} \Rightarrow \frac{dy}{dx} = 6x^{1/2} + x^{-1/2} + \frac{5}{2}x^{-3/2}$

2 Tangents and Normals

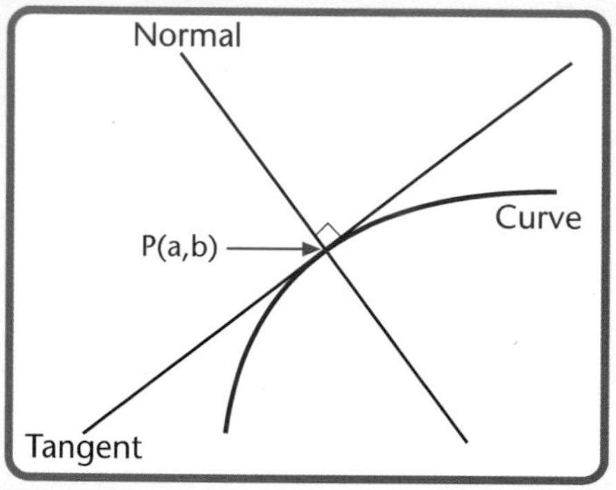

The *tangent (T)*, to the curve is a *line* touching the curve at the point, P.

The *normal (N)*, to the curve at P is a *line perpendicular* to the tangent at P.

In differentiation $\frac{dy}{dx}$ refers to the gradient of the tangent at a point.

Example 2

Find the equations of the tangent and the normal to the curve $y = x^3 - 3x^2 - 2x + 14$, at the point Q where $x = 2$.

Solution

The tangent and normal are lines ⇒ we need the gradient and a point.

- Point Q ⇒ $y = (2)^3 - 3(2)^2 - 2(2) + 14 = 6 \Rightarrow Q(2, 6)$
- Gradient of tangent ⇒ differentiate: $\frac{dy}{dx} = 3x^2 - 6x - 2$
- So at Q, gradient of T ⇒

  $\frac{dy}{dx} = 3(2)^2 - 6(2) - 2 = -2$
- Gradient and point ⇒ equation of T: $y - 6 = -2(x - 2) \Rightarrow y = -2x + 10$
- So, gradient of N = $\frac{-1}{-2} = \frac{1}{2}$
- Gradient and point ⇒ equation of N: $y - 6 = \frac{1}{2}(x - 2) \Rightarrow y = \frac{1}{2}x + 5$

Gradient Normal = $\dfrac{-1}{\text{Gradient Tangent}}$

3 Increasing and Decreasing Functions

A function is increasing where $\frac{dy}{dx} > 0$.

A function is decreasing where $\frac{dy}{dx} < 0$.

Example 3

Find the set of values of x, for which the curve with equation $y = x^3 - 4x^2 - 3x - 11$, is increasing.

Solution

- Differentiate: $\frac{dy}{dx} = 3x^2 - 8x - 3$
- Increasing function: $\frac{dy}{dx} > 0 \quad \Rightarrow 3x^2 - 8x - 3 > 0$
- Factorising: $(3x + 1)(x - 3) > 0$
- Number line:

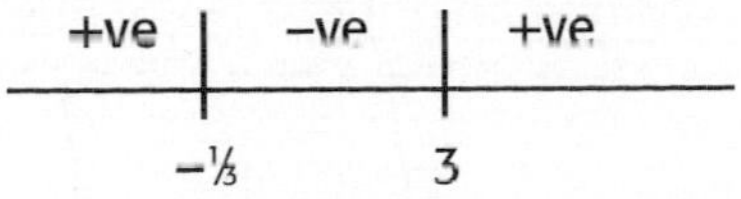

- Hence curve is increasing when $x < -1/3$ or $x > 3$

(Note that the curve is decreasing when $-1/3 < x < 3$)

4 Stationary Points

Stationary points (SPs) are points with coordinates such that $\frac{dy}{dx}$ (the gradient of the tangent at a point) is equal to zero. There are three types of stationary point: *maxima, minima* and point of *inflection.* We need to find the stationary point(s) and say what type they are.

To begin, we differentiate the function $y = f(x)$, setting the result equal to zero and find the value(s) for x when $\frac{dy}{dx} = 0$. If we are asked for the coordinates of the SP, we put the value of x back into the original equation to find the corresponding value of y.

DAY 1 2 3 4 5 6 7

To find the type of SP, we can use one of two tests:

Test 1: Gradient Test: Look at the gradient $\frac{dy}{dx}$, either side of the SP (e.g. if $x = 5$ at the SP, then we would work out the value of $\frac{dy}{dx}$ at $x = 4.9$ and $x = 5.1$, say).

We find out whether the gradients at these two points are positive or negative, and by looking at the table below, we can ascertain the type of SP.

Test 2: Second Derivative Test: Differentiating $\frac{dy}{dx}$ again, we get $\frac{d^2y}{dx^2}$. Then we insert the x value of the SP into $\frac{d^2y}{dx^2}$. If the answer is negative, it's a maximum SP; if positive it's a minimum SP. However, if the answer is zero, the SP could still be any of the three SPs, so then we have to perform the gradient test.

	Maximum	Minimum	Inflection
Shape	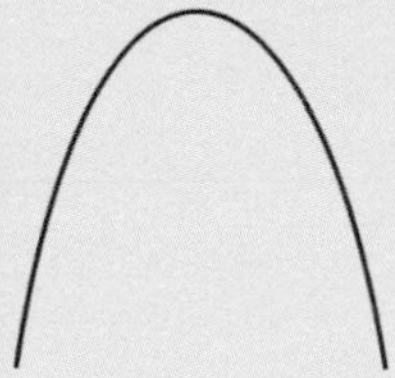	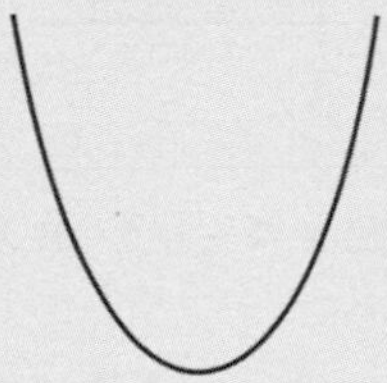	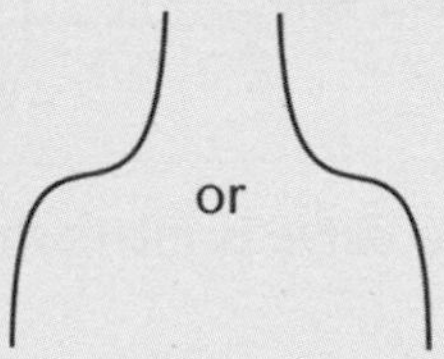
Gradient Test: $\frac{dy}{dx}$ is:	 + 0 −	 − 0 +	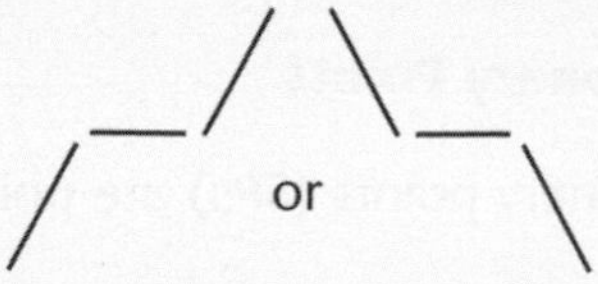 + 0 + − 0 −
2nd Derivative Test: $\frac{d^2y}{dx^2}$ is:	≤ 0	≥ 0	$= 0$

Example 4

For the curve $y = x^3 + 3x^2 - 9x + 15$, find:

a) the coordinates where the curve crosses the y-axis

b) coordinates of the two stationary points

c) nature of the stationary points.

Hence give a sketch of the curve.

d) State the set of values for x for which y is decreasing.

Solution

a) Crosses y-axis when $x = 0$, so $y = 15 \Rightarrow (0, 15)$

b) Differentiating: $\frac{dy}{dx} = 3x^2 + 6x - 9$

- Set $\frac{dy}{dx} = 0$ and solve for x: $3x^2 + 6x - 9 = 0 \Rightarrow 3(x^2 + 2x - 3) = 0$
  $\Rightarrow 3(x + 3)(x - 1) = 0 \Rightarrow x = -3$ or $x = 1$
- Find y-coordinates : $y = (-3)^3 + 3(-3)^2 - 9(-3) + 15 = 42$
  $y = (1)^3 + 3(1)^2 - 9(1) + 15 = 10$
- Hence the coordinates of the SPs are $(-3, 42)$ and $(1, 10)$.

c) Find the second derivative: $\frac{d^2y}{dx^2} = 6x + 6$

- When $x = -3$: $\frac{d^2y}{dx^2} = 6(-3) + 6 = -12 < 0 \Rightarrow$ Maximum SP at $(-3, 42)$
- When $x = 1$: $\frac{d^2y}{dx^2} = 6(1) + 6 = 12 > 0 \Rightarrow$ Minimum SP at $(1, 10)$

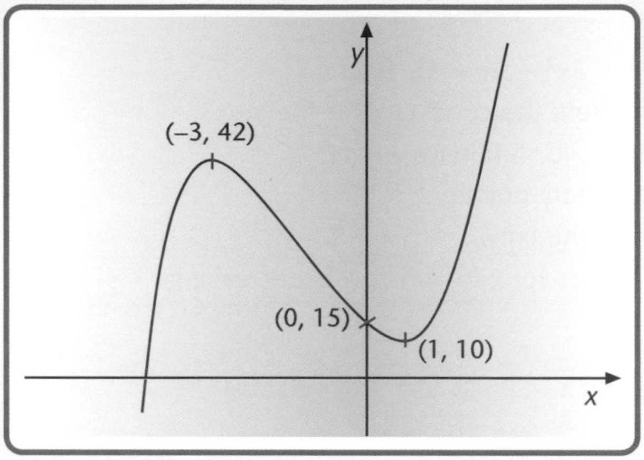

d) Looking at the graph, the curve is decreasing when: $-3 < x < 1$

Example 5

A Practical Problem: A cylindrical open cup is to be made from a thin sheet of silver of area 48π cm^2 with no wastage. The volume of the cup is V cm^3 and the radius is r cm.

a) Show that $V = \dfrac{\pi . r}{2}(48 - r^2)$.

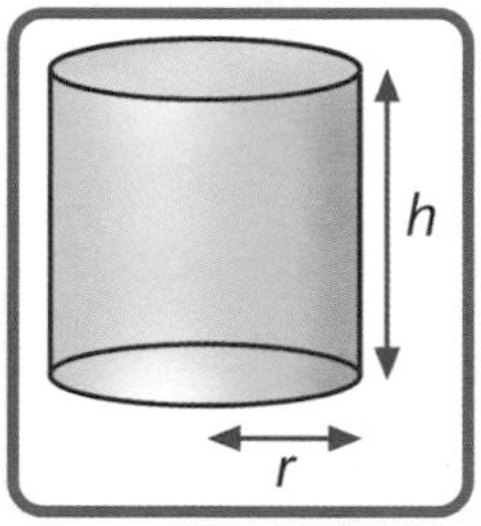

b) Find the value of r that gives a maximum volume.
c) Prove the volume is maximum.
d) Work out the maximum volume of the cup.

Solution

- Draw a picture to help.
- Write down formulae using the information given. We see that the question talks about volumes and surface areas, hence:

a) For a cylinder, the volume, $V = \pi r^2 h$ **(1)**

With no lid, surface area = curved bit + base $\Rightarrow 48\pi = 2\pi rh + \pi r^2$ **(2)**

- Look at the answer – we see that the answer is in terms of r, but the variable h does not appear, implying that it is our mission to eliminate h.

Hence $\dfrac{(2)}{\pi}$ gives $48 = 2rh + r^2 \Rightarrow h = \dfrac{48 - r^2}{2r}$ **(3)**

Substituting **(3)** into **(1)** gives: $V = \pi r^2 \left(\dfrac{48 - r^2}{2r}\right)$

After cancelling an *r*: $V = \dfrac{\pi . r}{2}(48 - r^2)$, as required.

b) $V = \dfrac{48\pi . r}{2} - \dfrac{\pi . r^3}{2}$, from a) $\Rightarrow \dfrac{dV}{dr} = \dfrac{48\pi}{2} - \dfrac{3\pi . r^2}{2} = 0$

Hence $\dfrac{48\pi}{2} = \dfrac{3\pi . r^2}{2} \Rightarrow r^2 = 16 \Rightarrow r = \pm 4$.

Since the radius cannot be negative, we accept $r = 4$ cm.

c) EITHER *Gradient Test:* Look at $\dfrac{dV}{dr}$ when $r = 3.9$ and $r = 4.1$

When $r = 3.9$, $\dfrac{dV}{dr} = \dfrac{48\pi}{2} - \dfrac{3\pi(3.9)^2}{2} = 3.723.... \rangle\ 0$

When $r = 4.1$, $\dfrac{dV}{dr} = \dfrac{48\pi}{2} - \dfrac{3\pi(4.1)^2}{2} = -3.817.... \langle\ 0$.

Hence 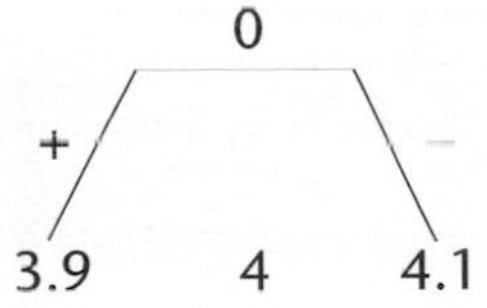

$\Rightarrow$ maximum volume when $r = 4$ cm.

OR *Second Derivative Test:* $\dfrac{d^2V}{dr^2} = -\dfrac{6\pi . r}{2} = -3\pi . r$

When $r = 4$, $\dfrac{d^2V}{dr^2} = -12\pi < 0 \Rightarrow$ maximum volume when $r = 4$ cm.

d) Maximum volume occurs when $r = 4$ cm, so substitute $r = 4$ in a).

$V = \dfrac{\pi . r}{2}(48 - r^2) = \dfrac{\pi . 4}{2}(48 - 16) = 64\pi \text{ cm}^3$

25 mins

Time Yourself

Differentiation

Have you improved?

1 a) Find the equation of the normal to the curve $y = \frac{x^2}{6} - x + \frac{1}{3}$, at the point P, where $x = 2$.

b) Hence find the coordinates of the point Q, where the normal meets the curve again.

1a) Find the coordinates of P. Find gradient of normal. Then find line

1b) Put normal = curve. Solve quadratic equation. Then find Q.

2 The tangent to the curve $y = 4x - kx^2$, (where k is a constant), at the point P where $x = 1$ cuts the y-axis at $y = 5$. Find:

a) the value of k

b) the equation of the tangent to the curve at P.

2a) Find point, gradient, then tangent in terms of k. Cuts y-axis at $x = 0$

2b) Use value of k to help!

3 A desk tidy with no lid is made from thin card into the form of a cuboid, as shown. The length of the box must be twice the width, where the width is x cm. The volume of the box is 1728 cm^3.

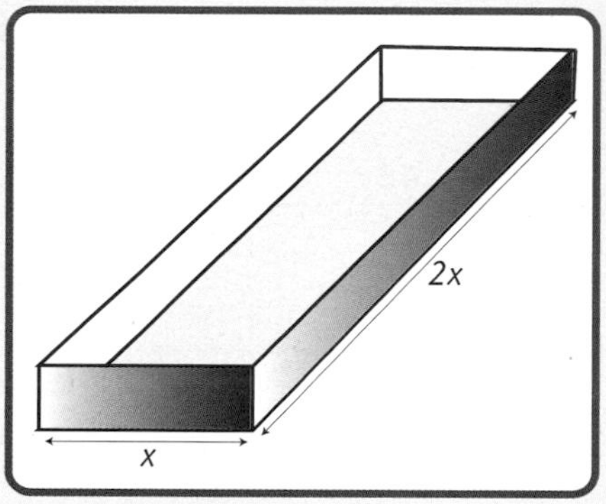

a) Show that the area of card, y cm^2, used to make the tray, is given by the equation: $y = \frac{5184}{x} + 2x^2$.

b) Given that x varies find the value of x, to 3 significant figures, for which y is least and prove that it is a minimum.

c) Deduce, to 3 significant figures, the minimum area of the card used to make the desk tidy.

3a) Let h = height. Write equations for V and y. Eliminate h

3b) Differentiate y and set to zero. Solve for x. Then apply test

3c) Substitute x value into 3a

DAY 1 2 3 4 5 6 7

Integration

60 mins
Time Yourself

How much do you know?

1 Find:

a) $\int 3-2x\,dx$

b) $\int 2x^3+5x^2-6x+3\,dx$

c) $\int (2x-3)(5-x)\,dx$

d) $\int (5+x)^2\,dx$

e) $\int 2\sqrt{x}+4\,dx$

f) $\int \frac{8}{5x^3}\,dx$

g) $\int \frac{4x^5-3x^3+2}{x^2}\,dx$

h) $\int \frac{2x^2+3x-1}{\sqrt{x}}\,dx$

2 The gradient of a curve is given by the equation $\frac{dy}{dx}=2x-3\sqrt{x}+3$, and passes through the point (4, 5).
Find the equation of the curve.

3 The region R is bounded by the curve, $y = x^2 + 3$, the x-axis, the y-axis, and the line, n, with equation $y = -3x + 13$.

a) Find the coordinates where the curve intersects the line n.

b) Draw a sketch of the situation, shading in the region R, required.

c) Hence calculate the area of the region R.

4 Find the area between the curve $y = 2x^2 - 7x$ and the x-axis.

Answers

1a) $3x - x^2 + c$ **b)** $\frac{1}{2}x^4+\frac{5}{3}x^3-3x^2+3x+c$ **c)** $-\frac{2}{3}x^3+\frac{13}{2}x^2-15x+c$

d) $25x+5x^2+\frac{1}{3}x^3+c$ **e)** $\frac{4}{3}x^{3/2}+4x+c$ **f)** $-\frac{4}{5}x^{-2}+c$ **g)** $x^4-\frac{3}{2}x^2-2x^{-1}+c$

h) $\frac{4}{5}x^{5/2}+2x^{3/2}-2x^{1/2}+c$ **2** $y=x^2-2x^{3/2}+3x-7$

3a) (−5, 28) & (2, 7) **b)** see diagram **c)** $16\frac{5}{6}$ (units)^2 **4** $14\frac{7}{24}$ (units)^2

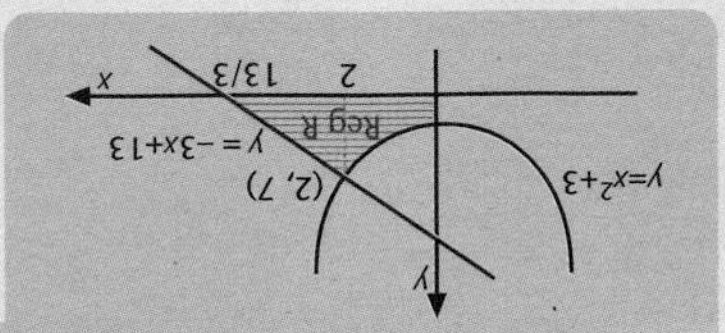

If you got them all right, skip to page 30

DAY 3

Integration

Learn the key facts

1 Basic Integration

Integration is the reverse process of differentiation. In AS Level mathematics, you are expected to integrate the following type of expression:

$$\int x^n dx = \frac{x^{n+1}}{n+1} + c, n \neq -1$$

Add one to the power, and divide by the new power!!!

Do not get worried about the ∫ and the dx; these just tell you to integrate! For indefinite integration (that's integration without limits) it is important that you include a constant **+** *c* at the end of your answer.

Example 1

The first three are fairly straightforward:

a) $\int x\,dx = \frac{x^2}{2} + c$ b) $\int 5\,dx = 5x + c$ c) $\int 3x - 4\,dx = \frac{3x^2}{2} - 4x + c$

Sometimes algebraic manipulation is needed before we can integrate:

d) $\int (2x+5)(x-3)dx$

$= \int 2x^2 - x - 15\,dx$

$= \frac{2}{3}x^3 - \frac{x^2}{2} - 15x + c$

e) $\int (3-x)^2\,dx$

$= \int 9 - 6x + x^2\,dx$

$= 9x - 3x^2 + \frac{x^3}{3} + c$

f) $\int \frac{3}{4x^2}dx = \int \frac{3}{4}x^{-2}\,dx$

$\frac{3}{4}.\frac{x^{-1}}{-1} + c = -\frac{3}{4x} + c$

g) $\int 4.\sqrt[3]{x}\,dx = \int 4.x^{1/3}\,dx$

$= \frac{4x^{4/3}}{4/3} + c = 3.x^{4/3} + c$

h) $\int \frac{6x^3 + 3x^2 - 4}{2x^2}\,dx$

$= \int \frac{6x^3}{2x^2} + \frac{3x^2}{2x^2} - \frac{4}{2x^2}\,dx$

$= \int 3x + \frac{3}{2} - 2x^{-2}\,dx$

$= \frac{3}{2}x^2 + \frac{3}{2}x + 2x^{-1} + c$

i) $\int \frac{x^2 + 3x + 5}{\sqrt{x}}dx$

$= \int \frac{x^2}{x^{1/2}} + \frac{3x}{x^{1/2}} + \frac{5}{x^{1/2}}dx$

$= \int x^{3/2} + 3x^{1/2} + 5x^{-1/2}\,dx$

$= \frac{2}{5}x^{5/2} + 2x^{3/2} + 10x^{1/2} + c$

2 The Constant of Integration

In some questions you will need to find the value of c, the constant of integration.

Example 2

The gradient of the curve is given by the equation, $\frac{dy}{dx} = 3x^2 - 6x + 5$, and the curve passes through the point (1, 10).
Calculate the equation of the curve.

Solution

We are given the equation for the gradient, $\frac{dy}{dx}$; we want the curve '$y = ...$'.

To go from back to '$y = ...$', we integrate:

- $y = \int 3x^2 - 6x + 5\ dx = x^3 - 3x^2 + 5x + c$, and we know when $x = 1$, $y = 10$,
- So $10 = 1^3 - 3(1)^2 + 5(1) + c \Rightarrow 10 = 3 + c \Rightarrow c = 7$.
- Hence the equation of the curve is: $y = x^3 - 3x^2 + 5x + 7$.

3 Definite Integration and Finding Regions Under Curves

Definite integration is integration involving the use of limits.

Example 3

Evaluate the following integral: $\int_1^4 3x^2 - 3\sqrt{x}\ dx$

Solution

- Integrate as usual; do not use + C $\qquad \int_1^4 3x^2 - 3x^{1/2}\ dx$
- Put integrand in square brackets with limits $\qquad = \left[x^3 - 2x^{3/2}\right]_1^4$
- Put in the limits and work out 'top expression' – 'bottom expression' $\qquad = (4^3 - 2(4)^{3/2}) - (1^3 - 2(1)^{3/2})$ $\quad = (64 - 16) - (1 - 2)$
- Evaluate $\qquad = 49$

One application of integration is to find area between curves and the coordinate axes.

Example 4

$$\int_1^4 x^2 dx = \left[\frac{x^3}{3}\right]_1^4 = \left(\frac{64}{3} - \frac{1}{3}\right) = 21 \text{ (units)}^2$$

This represents the area under the curve $y = x^2$, between the lines $x = 1$ and $x = 4$, as shown in the diagram. Hence definite integration with respect to x can represent the area between the curve and the x-axis.

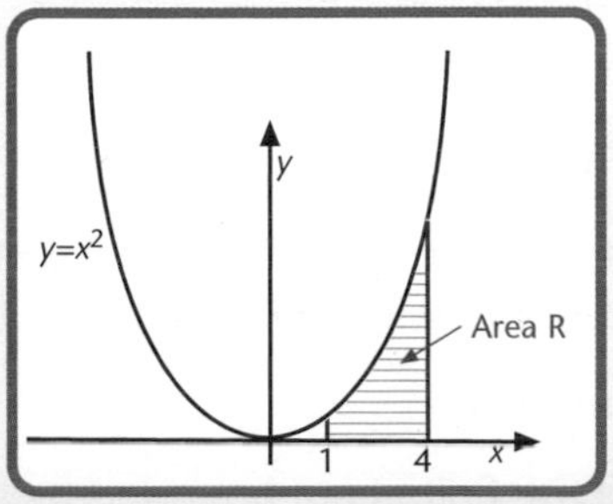

DAY 1 2 3 4 5 6 7

Example 5

The region R is bounded by the curve $y = x^2 + 4$ and the line l with equation $y = -2x + 12$.

a) Find the coordinates where the curve intersects the line l.

b) Hence calculate the area of the region R.

Solution

a) Curve intersects ⇒ curve = line ⇒ solve equations simultaneously.

- Hence $x^2 + 4 = -2x + 12$, giving $x^2 + 2x - 8 = 0$
  $\Rightarrow (x + 4)(x - 2) = 0 \Rightarrow x = 2$ or $x = -4$
- When $x = 2$, $y = 2^2 + 4 = 8$ and when $x = -4$, $y = (-4)^2 + 4 = 20$
- Coordinates of intersection are (2, 8) and (–4, 20)

b) To help us to find the area of R, we usually draw a sketch to see what is happening.

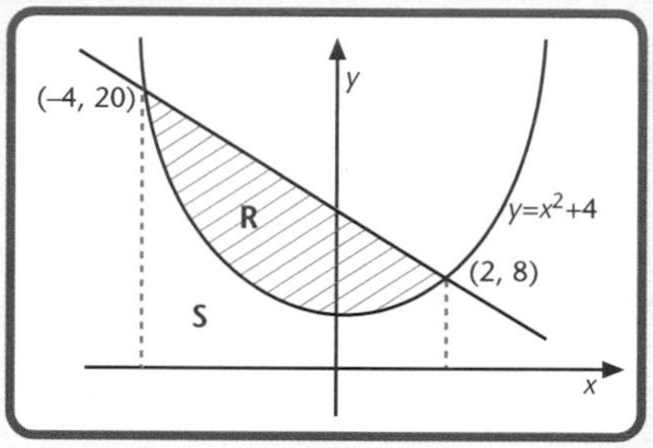

Curve of x^2 is moved up 4 units

Line has negative gradient and must intersect the curve twice

Region R is between curve and line

The region S, has been shown in to help us.

The regions R and S together form a quadrilateral known as a trapezium.

Area of trapezium $= \frac{1}{2}(a + b)h$

Area (R + S) $= \frac{1}{2}(20 + 8)(6) = 84$ (units)^2

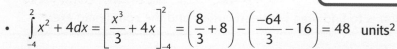

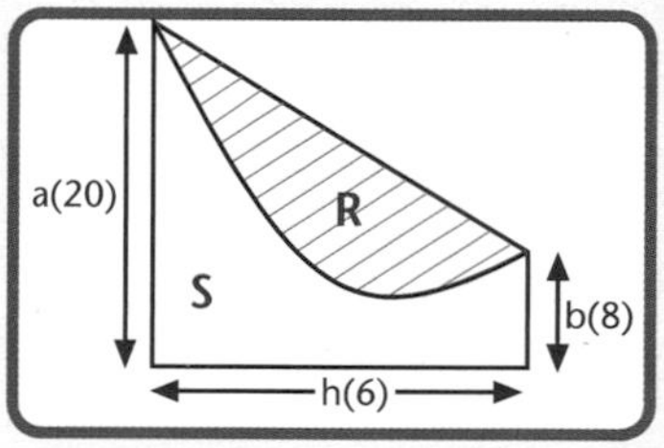

If we subtract region S from the trapezium, we are left with region R.
S is found by integrating the curve $y = x^2 + 4$ between $x = -4$ and $x = 2$.

- $\int_{-4}^{2} x^2 + 4dx = \left[\frac{x^3}{3} + 4x\right]_{-4}^{2} = \left(\frac{8}{3} + 8\right) - \left(\frac{-64}{3} - 16\right) = 48$ units^2

- Hence: area (R) = area(R + S) – area(S) = 84 – 48 = 36 units^2.

Finding Regions Below the x-axis

Integration will give us a negative value if we integrate a region below the x-axis.

Example 6

Find the finite area bounded by the curve $y = x^2 - 6x$ and the x-axis.

Solution

- $y = x(x - 6) = 0 \Rightarrow x = 0$ & $x = 6$, where the curve cuts the x-axis.

It's a happy curve!

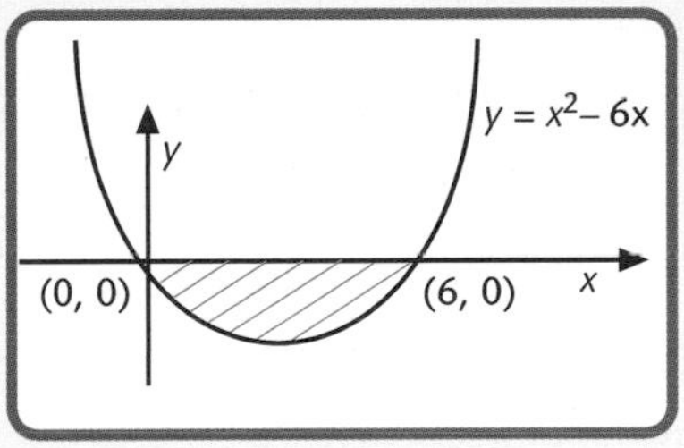

Area required $= \int_0^6 (x^2 - 6x)dx$

$$= \left[\frac{x^3}{3} - 3x^2\right]_0^6 = \left(\frac{216}{3} - 108\right) - (0)$$

$= -36$, which is negative!

Since areas must be positive we take the modulus: area = 36 (units)^2

Example 7

The region S is bounded by the curve with equation $y = x^2 - 4x - 12$ and the line l with equation $y = 3x - 18$. Calculate the area of the region S.

Solution

- Find the coordinates of intersection $\Rightarrow$ curve = line (again!)
- Hence $x^2 - 4x - 12 = 3x - 18$, which gives $x^2 - 7x + 6 = 0$
  $\Rightarrow (x - 6)(x - 1) = 0 \Rightarrow x = 6$ or $x = 1$
- When $x = 6$, $y = 3(6) - 18 = 0$ and when $x = 1$, $y = 3(1) - 18 = -15$
- Coordinates of intersection are (6, 0) and (1, −15)

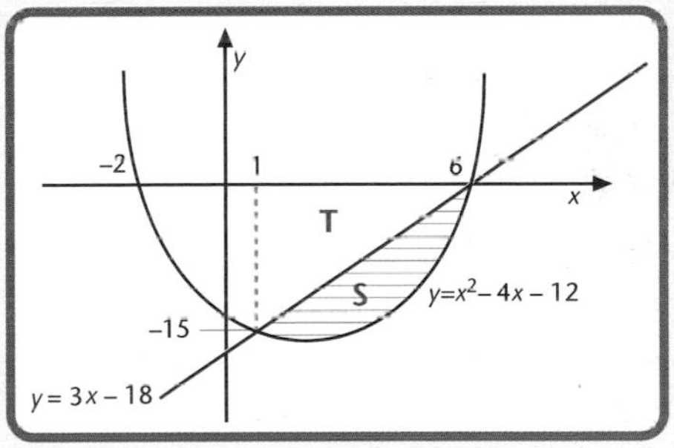

Again, we need to sketch the curve to find out the region that we want:

Curve: $y = x^2 - 4x - 12$
$= (x - 6)(x + 2)$
is a 'happy' quadratic curve cutting the x-axis at $x = -2$ and $x = 6$.

- $\int_1^6 x^2 - 4x - 12\, dx = \left[\frac{x^3}{3} - 2x^2 - 12x\right]_1^6 = (72 - 72 - 72) - (\frac{1}{3} - 2 - 12) = -58\frac{1}{3} < 0$

- So area (S + T) = $-1 \times -58\frac{1}{3} = 58\frac{1}{3}$(units)^2

- Since region T is a triangle: area (T) = $\frac{1}{2} b.h = \frac{1}{2}(5)(15) = 37\frac{1}{2}$ (units)^2

- $\therefore$ area (S) = area (S + T) − area (T) = $58\frac{1}{3} - 37\frac{1}{2} = 20\frac{5}{6}$(units)^2

Integration

Have you improved?

1 Given that $f(x) = \dfrac{(2x+1)(2x-1)}{\sqrt{x}}$; $x > 0$

a) Show that $f(x)$ can be expressed in the form
$f(x) = Ax^{3/2} + Bx^{-1/2}$, where A and B are integers to be determined.

b) Show that $\int_1^2 f(x)\,dx = \frac{2}{5}(C\sqrt{2} + D)$, stating the values of the integers C and D.

1a) Multiply out the top. Split up the resulting fraction.

1b) Use form of f(x) that you have found in a). NB: $\sqrt{2}\ \sqrt{2} = 2$

2

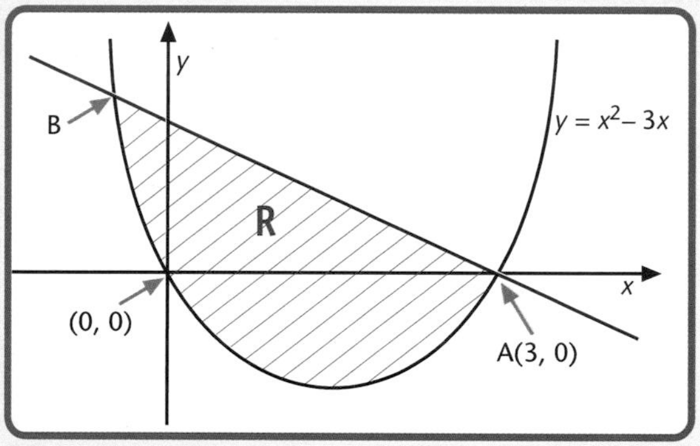

Figure 1

Figure 1 shows a sketch of the curve $y = x^2 - 3x$, and the line AB that is the normal to the curve at the point A(3, 0).
The finite region R is bounded by the curve and the normal as shown in Figure 1.

a) Find the equation of the normal to the curve at the point A(3, 0).
b) Calculate the coordinates of the point B, where the normal meets the curve again.
c) Find, to 3 significant figures, the area of the region R.

2a) Use differentiation. Gradient and point!

2b) Put 'curve = normal'. Solve quadratic eqn

2c) You need to split R up into various regions

Trigonometry

30 mins
Time Yourself

How much do you know?

1 Sketch the graphs of:
a) $y = \cos x$ b) $y = \tan x$ c) $y = \sin 2x$ d) $y = \sin(x - 30) + 2$

2 a) A is an acute angle and $\cos A = \frac{12}{13}$. Find $\sin A$ and $\tan A$

b) $\cos 53° = 0.6$. Use this information to find:
i) $\sin 53°$ ii) $\tan 37°$
c) Write down the values of:
i) $\cos 30°$ ii) $\tan 60°$, giving each answer in surd form.

3 Find the solutions to each of the following equations in the specified ranges:
a) $\tan\left(\frac{x}{2}\right) = 0.5;\ -360° < x < 360°$ b) $\cos(3x - 90) = -\frac{\sqrt{3}}{2};\ 0 \le x \le 270°$

4 Find the solution to the following equations in the specified ranges:
a) $4\sin^3 x - \sin x = 0;\ -180° \le x \le 180°$ b) $7\sin x + 6\cos^2 x = 8;\ 0 \le x \le 180°$
c) $2\sin x - 3\sin x\cos x = 0;\ 0 \le x \le 180°$ d) $2\sin x - 3\cos x = 0;\ 0 \le x \le 180°$

Answers

1 See graphs in section 1 of 'Learn the key facts'
2a) $\sin A = \frac{5}{13}$ $\tan A = \frac{5}{12}$ b) i) 0.8 ii) 0.75 c) i) $\frac{\sqrt{3}}{2}$ ii) $\sqrt{3}$
3a) $x = 53.1°, -306.9°$ b) $x = 80°, 200°, 100°, 220°$
4a) $x = 0°, 180°, -180°, 30°, 150°, -30°, -150°$ b) $x = 41.8°, 138.2°, 30°, 150°$
c) $x = 0°, 48.2°, 180°$ d) $x = 56.3°$

If you got them all right, skip to page 38

DAY 1 2 3 4 5 6 7

Trigonometry

Learn the key facts

1 Graphs of Trigonometric Functions

You need to be able to sketch the graphs of sin*x*, cos*x* and tan*x*. Key things to remember are shown below.

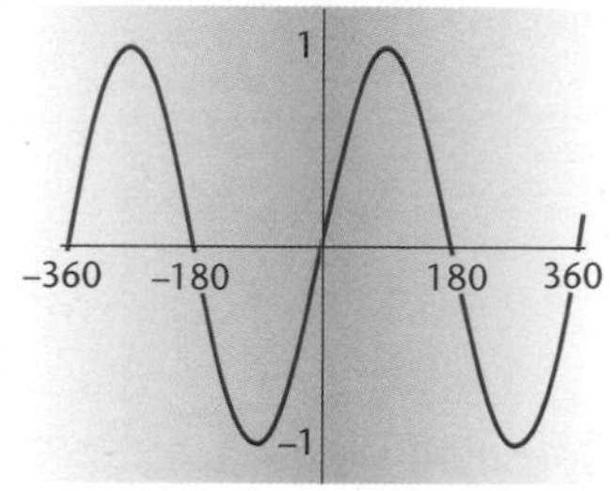

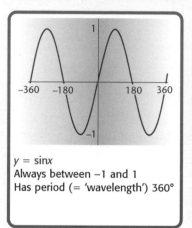

y = sin*x*
Always between –1 and 1
Has period (= 'wavelength') 360°

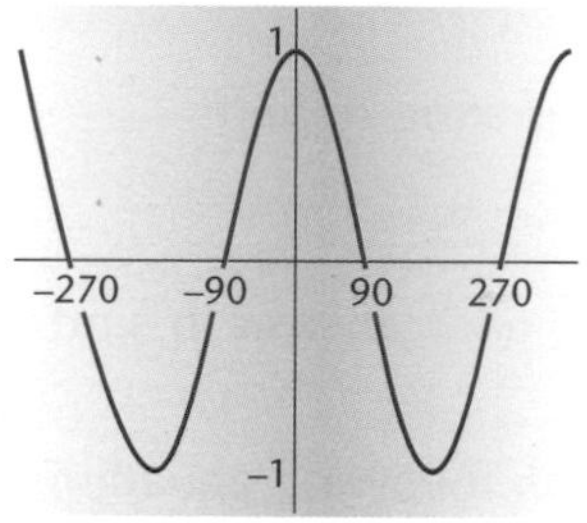

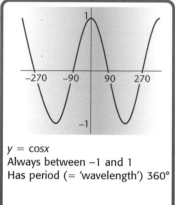

y = cos*x*
Always between –1 and 1
Has period (= 'wavelength') 360°

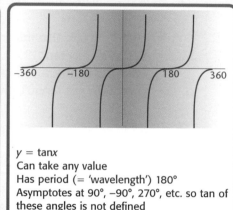

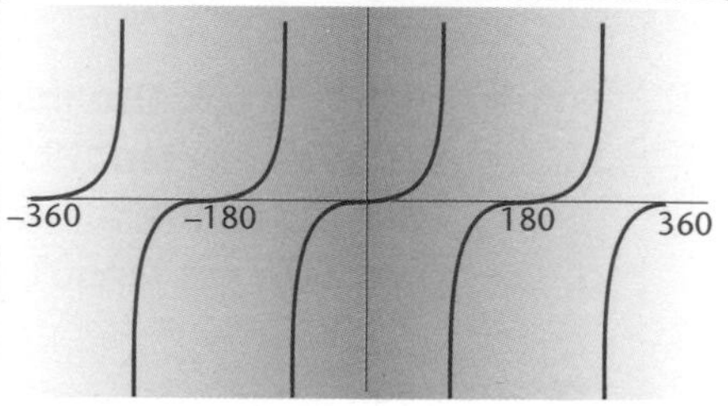

y = tan*x*
Can take any value
Has period (= 'wavelength') 180°
Asymptotes at 90°, –90°, 270°, etc. so tan of these angles is not defined

You can use these graphs to sketch others, using the rules for transformations of graphs from GCSE Higher.
Here's a reminder:

- Numbers outside the brackets affect the *y*-values. If the number is added or subtracted, the graph is shifted up or down. If you are multiplying by the number, the graph is stretched in the *y* direction
- Numbers inside the brackets affect the *x*-values. Adding or subtracting still mean a shift, and multiplying still means a stretch, but the graph is affected the 'wrong' way –
  e.g. +2 inside the brackets results in a shift of the graph by 2 in the **negative** *x* direction;
  ×2 inside the brackets results in the graph being **squashed** by a factor of 2 in the *x* direction.
  So if you had **sin2*x***, the period would be **halved**. For **cos(*x*/3)**, the period would be **trebled**.
- A minus outside the brackets reflects the graph in the *x*-axis; minus inside reflects it in the *y*-axis.

Some examples of using these follow.

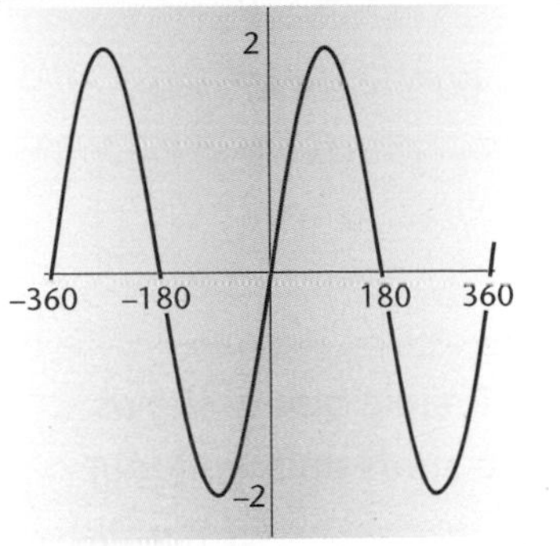

This is $y = 2\sin x$
The original graph has been stretched by a factor of 2 in the y direction

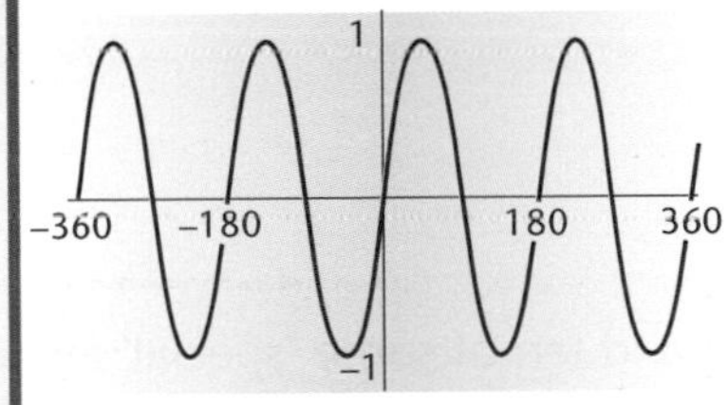

This is $y = \sin 2x$
The original graph has been squashed by a factor of 2 in the x direction

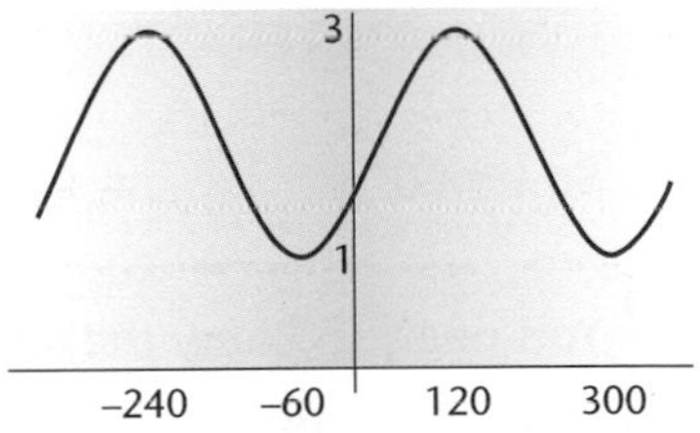

This is $y = \sin(x - 30) + 2$
The original graph has shifted 30 in the positive x direction and 2 in the positive y direction

2 Trig Ratios of Special Angles and Using sin to Find cos

If you are told the sin, cos or tan of an acute angle you can work out the other ones by drawing a triangle and using 'SOHCAHTOA' and Pythagoras.

Example 1

A is an acute angle and $\sin A = \frac{3}{5}$. Find $\cos A$ and $\tan A$.

Solution

We put the opposite as 3 and the hypotenuse as 5 because $\sin = \frac{\text{opp}}{\text{hyp}}$

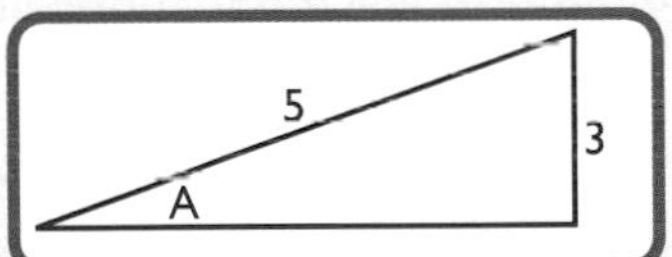

Using Pythagoras, we find the other side is 4.
So $\cos A = \frac{\text{adj}}{\text{hyp}} = \frac{4}{5}$ and $\tan A = \frac{\text{opp}}{\text{adj}} = \frac{3}{4}$

The following rules also enable you to work out trig ratios from ones you know:

$\tan x = \dfrac{\sin x}{\cos x}$ $\quad \sin(90 - x) = \cos x \quad \cos(90 - x) = \sin x \quad \sin^2 x + \cos^2 x = 1$

Learn!!

Example 2

$\sin 30° = \frac{1}{2}$. Use this information to find: a) $\cos 30°$ b) $\tan 60°$

Solution

a) $\cos 30° = \sqrt{1 - \sin^2 30} = \sqrt{1 - \frac{1}{4}} = \sqrt{\frac{3}{4}} = \frac{\sqrt{3}}{2}$

b) $\tan 60° = \dfrac{\sin 60°}{\cos 60°}$

But $\sin 60° = \cos(90° - 60°) = \cos 30° = \frac{\sqrt{3}}{2}$ and $\cos 60° = \sin(90° - 60°) = \sin 30° = \frac{1}{2}$

So $\tan 60° = \dfrac{\sqrt{3}}{2} \Big/ \dfrac{1}{2} = \sqrt{3}$

You need to know the sin, cos and tan of some 'special' angles. The table below shows a way of remembering them – but you must be able to simplify the surds and work out tan from sin and cos.

	0°	30°	45°	60°	90°
sin	$\sqrt{\frac{0}{4}}$	$\sqrt{\frac{1}{4}}$	$\sqrt{\frac{2}{4}}$	$\sqrt{\frac{3}{4}}$	$\sqrt{\frac{4}{4}}$
cos	$\sqrt{\frac{4}{4}}$	$\sqrt{\frac{3}{4}}$	$\sqrt{\frac{2}{4}}$	$\sqrt{\frac{1}{4}}$	$\sqrt{\frac{0}{4}}$

3 Solving Simple Trigonometric Equations – CAST Diagrams

You are often required to solve equations like $\sin(2x - 15°) = \frac{1}{2}$ $\quad -360° < x < 360°$

A CAST diagram is used to do this. This diagram tells you where each of the three trig functions is positive. Example 3 shows the key steps in answering this sort of question.

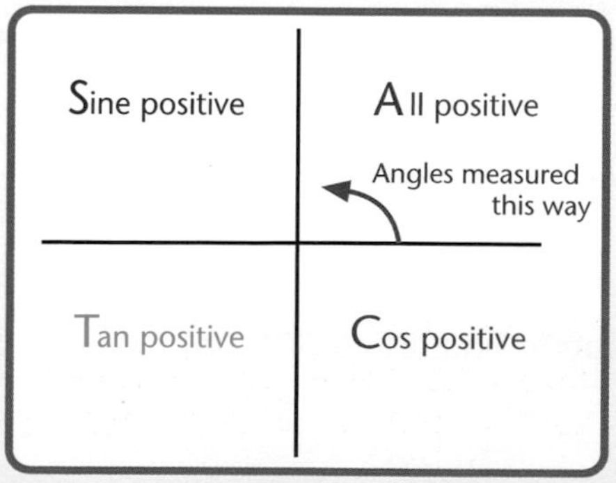

Example 3

Find all solutions to the following equations in the specified ranges:

a) $\sin 2x = \frac{1}{2}$; $0° < x° < 360°$ b) $\cos(x - 45) = -\frac{1}{\sqrt{2}}$; $-270° \le x° \le 270°$

Solution

a) STEP 1: Set $2x = u \Rightarrow$ we are solving $\sin u = \frac{1}{2}$

STEP 2: Find the values that u must be between.
$0° < x° < 360° \Rightarrow 0° < 2x < 720°$

STEP 3: Decide which quadrants we want
$\Rightarrow$ since $\sin u$ is positive, we want 1st and 2nd

STEP 4: Use the calculator (or 'special' angles) to find $\sin^{-1}\frac{1}{2} = 30°$

STEP 5: Use the diagram to find 2nd solution; so the two solutions are 30° and 150°

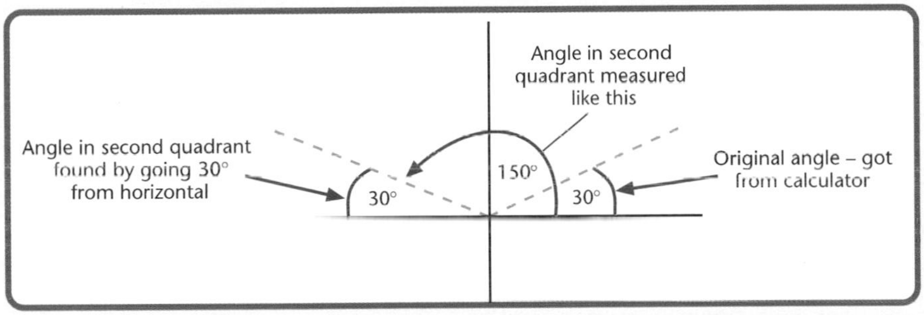

Always measure from the horizontal

STEP 6: Add and/or subtract 360° to these until we are outside range for u.
$\Rightarrow$ From 30°, get 390°, 750° (outside), −330° (outside)
$\Rightarrow$ From 150° get 510°, 870° (outside), −210°(outside)

STEP 7: Find the values for $x \Rightarrow$ We have $u = 30°, 390°, 150°, 510°$
$\Rightarrow$ So $2x = 30°, 390°, 150°, 510° \Rightarrow x = 15°, 195°, 75°, 255°$

b) 1) $u = x - 45$

2) $-270° \le x° \le 270° \Rightarrow -315° \le u° \le 225°$

3) cos is negative – so 2nd and 3rd quadrants

4) $\cos^{-1} -\frac{1}{\sqrt{2}} = 135°$

5) solutions are 135° and 225°

6) 135° gives 495° (outside), −225°; 225° gives 585° (outside), −135°

7) $x - 45 = u = 135°, -225°, 225°, -135° \Rightarrow x = 180°, -180°, 270°, -90°$

4 Trig Equations using Quadratics or Cubics and Trig Rules

Polynomials

Sometimes you will meet equations looking like $2\sin^2 x - \sin x - 1 = 0$ – in other words, quadratics (or maybe cubics) with a trig function in them instead of just x. These are easy – you just factorise (put $y = \sin x$ if it makes it easier for you), get the values for $\sin x$, then get the solutions as above.

Example 4

Solve the equation $2\sin^2 x - \sin x - 1 = 0;\ 0° < x° < 360°$

Solution

We can change this to $2y^2 - y - 1 = 0 \Rightarrow (2y + 1)(y - 1) = 0 \Rightarrow y = -\frac{1}{2}$ or 1

$\Rightarrow \sin x = -\frac{1}{2}$ or 1

$\sin x = -\frac{1}{2} \Rightarrow x = 210°, 330°;\ \sin x = 1 \Rightarrow x = 90°$

Using Trig Rules

Some equations need you to use the rules you were given in section 2.

$\sin^2 x + \cos^2 x = 1$ and $\tan x = \dfrac{\sin x}{\cos x}$ are the most useful here!

- You use $\sin^2 x + \cos^2 x = 1$ if you have both sin and cos in the equation and at least one is squared.
- You are aiming to get the whole equation in terms of just sin or just cos.
- You use $\tan x = \dfrac{\sin x}{\cos x}$ if you end up with number × sin = number × cos.
- If you have anything else – try getting it all one side and factorise.
- NEVER 'cancel' something like $\sin x$ from both sides of an equation. Take it out as a common factor instead.

Example 5

Solve the following equations for $0° \leq x \leq 180°$

a) $\sin 3x = 4\cos 3x$ b) $2\sin^2 2x + \cos 2x = 1$ c) $2\sin x\cos x = \sin x$

Solution

a) We can't use $\sin^2 x + \cos^2 x = 1$ because nothing is squared.

It is in the form number × sin = number × cos, so try using $\tan = \frac{\sin}{\cos}$

Dividing both sides of the equation by $\cos 3x$, we get: $\frac{\sin 3x}{\cos 3x} = 4 \Rightarrow \tan 3x = 4$

$\Rightarrow 3x = 76°, 256°, 436° \Rightarrow x = 25.3°, 85.3°, 145.3°$

b) As we've got a $\sin^2$, try using $\sin^2 + \cos^2 = 1$

So $\sin^2 2x = 1 - \cos^2 2x \Rightarrow 2(1 - \cos^2 2x) + \cos 2x = 1 \Rightarrow -2\cos^2 2x + \cos 2x + 1 = 0$

This is a quadratic: factorising gives $(1 + 2\cos 2x)(1 - \cos 2x) = 0$,

giving $\cos 2x = -\frac{1}{2}, 1$

$\Rightarrow 2x = 120°, 240°, 0°, 360° \Rightarrow x = 60°, 120°, 0°, 180°$

c) This isn't in either of the forms we recognise – so try factorising!

$2\sin x\cos x - \sin x = 0 \Rightarrow \sin x(2\cos x - 1) = 0$ so $\sin x = 0$ or $\cos x = \frac{1}{2}$

So $x = 0°, 60°, 180°$

Trigonometry

Have you improved?

1 a) Sketch on the same axes the graphs of $y = \sin 2x$ and $y = \cos 2x$ for $-180° \leq x \leq 180°$

b) Hence state the number of solutions of the equation $\tan 2x = 1$; $-180° \leq x \leq 180°$

2 Solve the equation $2\sin x\cos x - \sin x - 2\cos x + 1 = 0$ for $0° \leq x \leq 360°$

3 Solve the equation $\tan 2x + 2\sin 2x = 0$; $-180° \leq x \leq 180°$

4 $\sin A = \frac{1}{7}$

Find the exact value of:

a) $\cos A$ b) $\cos(90° - A)$ c) $\sin(180° - A)$ d) $\cos(180° + A)$

5 Sketch the graph of $y = \tan(2x - 60)°$ for $-90° \leq x \leq 90°$, showing clearly the positions of asymptotes and the points at which the graph crosses the coordinate axes.

1a) Remember the period will be halved

1b) Write tan in terms of sin and cos. Sin2x and cos2x are equal when the graphs cross

2) This will factorise – replace sinx and cosx by s and c if it makes it easier

3) Write tan in terms of sin and cos. Cross multiply and factorise

4a,b) Look back at the rules in section 2

4c,d) Use a CAST diagram to tell you how these relate to sinA and cosA

5) 2x – 60 = 2(x – 30). Work out what tan(x – 30) looks like first

Series

25 mins
Time Yourself

How much do you know?

1 a) The ninth term and the twenty-ninth term of an arithmetic series are 9 and −3 respectively. Find:

i) the first term and common difference of the series

ii) the sum of the first thirty terms of the series.

b) The eighth, ninth and tenth terms of an arithmetic progression are $4y - 2$, $3y + 12$ and $5y - 1$ respectively. Find:

i) the value of y

ii) the common difference and the first term of the series.

2 The third and fourth terms of a geometric progression are 21 and 14 respectively. Find:

a) the common multiple and the first term of the series

b) the sum of the first 15 terms of the series

c) the difference between the sum to infinity and the sum of the first 15 terms.

3 Evaluate the following expressions, giving your answers to 3 significant figures:

a) $\sum_{r=1}^{15} 4r - 3$ b) $\sum_{r=5}^{18} 5(0.75)^r$ c) $\sum_{r=1}^{12} 1.3^r + 2.5r$

Answers

1a) i) $a = 13.8$, $d = -0.6$ ii) 153 b) i) 9 ii) $d = 5$, $a = -1$
2a) $r = 2/3$, $a = 47.25$ b) 141.42629... c) $S_\infty = 141.75 \Rightarrow$ difference = 0.324(3dp) **3a)** 435 b) 4.66 c) 292

If you got them all right, skip to page 44

Spend no more than 35 mins on this topic

Series

Learn the key facts

There are two types of series that you need to know. They are an *arithmetic series* and a *geometric series.*

1 Arithmetic Series

An arithmetic series (or progression) has a constant difference d. Examples are:
$3 + 5 + 7 + 9 + 11 + \ldots$ where $d = 2$, and
$27 + 23 + 19 + 15 + 11 + \ldots$ where $d = -4$.
You need to learn how to use the following formulae that are associated with arithmetic series:

nth term $\equiv a + (n - 1)d$ **(1)**

Sum to n terms, $S_n = \frac{n}{2}\{2a + (n-1)d\}$ *or* $S_n = \frac{n}{2}(a + l)$ **(2)**

For three consecutive terms: x, y and z, then $y - x = z - y$ $(= d)$ **(3)**
where: a is the first term, d is the common difference, l is the last term of the series.

For example: For the arithmetic series: $4 + 7 + 10 + 13 + \underline{16} + 19 + 22 + \ldots$
$a = 4$, $d = 3$ (i.e. $7 - 4 = 3$, $10 - 7 = 3$, $13 - 10 = 3$, etc.)
The fifth term (underlined) of the series is: $4 + (5 - 1)(3) = 4 + 12 = \underline{16}$ using **(1)**

Sum of the first 6 terms is: $S_6 = \frac{6}{2}\{2.4 + (6-1).3\} = 3(8 + 15) = 69$ using **(2)**

Example 1
The ninth term of an arithmetic series is 7 and the sum of the first twenty-five terms is 325. Find:
a) the common difference and
b) the first term of the series.

Solution
a) 9th term is 7: $a + (9 - 1)d = 7$, i.e. $a + 8d = 7$ **(4)**
- $S_{25} = 325$: $\frac{25}{2}(2a + (25 - 1)d) = 325$, i.e. $12.5(2a + 24d) = 325$ **(5)**
- Simultaneous equations: 2**(4)** $\Rightarrow 2a + 16d = 14$
- **(5)**/12.5 $\Rightarrow 2a + 24d = 26$
- Solving gives: $8d = 12 \Rightarrow d = 1.5$

b) Finding a: $a + 8(1.5) = 7 \Rightarrow a = 7 - 12 = -5$

DAY 4

2 Geometric Series

A geometric series (or progression) has a constant ratio (or multiple) called r. Examples are: $2 + 6 + 18 + 54 + \ldots$ where $r = 3$; and

$512 + 256 + 128 + 64 + \ldots$ where $r = \frac{1}{2}$.

You need to learn how to use the following formulae associated with geometric series:

nth term $\equiv ar^{n-1}$ **(6)**

Sum to n terms, $S_n = \dfrac{a(1-r^n)}{1-r}$ **(7)**

For three consecutive terms: x, y and z, then $\dfrac{y}{x} = \dfrac{z}{y}\ (= r)$ **(8)**

When the common ratio, r, is such that $-1 < r < 1$, then the sum of a geometric series converges to a finite value and has a sum to infinity, given by: $S_\infty = \dfrac{a}{1-r}$ **(9)**

Example 2

The third term and the sixth term of a geometric series are 8 and $-\frac{1}{8}$ respectively. Find: (a) the common ratio, (b) the first term and (c) the sum to infinity of the series.

Solution

a) 3rd term is 8: $ar^{3-1} = 8$, i.e. $ar^2 = 8$ **(10)**

- 6th term is $-\frac{1}{8}$: $ar^{6-1} = -\frac{1}{8}$, i.e. $ar^5 = -\frac{1}{8}$ **(11)**

- Divide **(11)** by **(10)**: $\dfrac{ar^5}{ar^2} = \dfrac{-\frac{1}{8}}{8} \Rightarrow r^3 = -\dfrac{1}{64} \Rightarrow r = -\frac{1}{4}$

b) Substituting: **(10)** $\Rightarrow a = \dfrac{8}{r^2} = \dfrac{8}{1/16} = 128$

c) Formula: $S_\infty = \dfrac{128}{1--(1/4)} = 102.4$

DAY 4

Example 3

The second, third and fourth term of a geometric series are x, $3x + 2$ and $11x + 10$ respectively where x is positive. Find: a) the value of x, b) the common multiple, c) the first term and d) the sum of the first 12 terms, giving your answer to 2 significant figures.

Solution

a) 3 consecutive terms $\Rightarrow \dfrac{3x+2}{x} = \dfrac{11x+10}{3x+2}\ (=r)$ using **(8)**

- Algebra: $(3x + 2)(3x + 2) = x(11x + 10) \Rightarrow 9x^2 + 12x + 4 = 11x^2 + 10x$
- Form into a quadratic: $2x^2 - 2x - 4 = 0 \Rightarrow 2(x - 2)(x + 1) = 0$
- Solving: $x = 2$ or $-1 \Rightarrow$ Since $x > 0$, then $x = 2$

b) Using **(8)**: $r = \dfrac{3x+2}{x} = \dfrac{3(2)+2}{2} = 4$

c) Second term is $ar = x$: $a(4) = 2$. So $a = \frac{1}{2}$.

d) Formula: $S_{12} = \dfrac{0.5(1-4^{12})}{1-4} = 2796202.5 = 2800000(2\text{sf})$

3 Use of Sigma (Σ) Notation

$\sum_{r=a}^{b} u_r$ represents the sum of a series whose first term is u_a; nth term is u_n and last term is u_b. Also the number of terms $= b - a + 1$

Example 4

Evaluate: $\sum_{r=3}^{30} 3r - 5 + 1.3^r$ giving your answer to 3 significant figures.

Solution

- Split up: $\sum_{r=3}^{30} 3r - 5 + 1.3^r = \sum_{r=3}^{30} 3r - 5 + \sum_{r=3}^{30} 1.3^r$

- $\sum_{r=3}^{30} 3r - 5 = (3(3) - 5) + (3(4) - 5) + (3(5) - 5) + (3(6) - 5) + \dots$

  $= 4 + 7 + 10 + 13 + \dots$

which is an AP: $a = 4$, $d = 7 - 4 = 3$, $n = b - a + 1 = 30 - 3 + 1 = 28$

- So: $\sum_{r=3}^{30} 3r - 5 = \frac{28}{2}(8 + 27(3)) = 1246$

- $\sum_{r=3}^{30} 1.3^r = 1.3^3 + 1.3^4 + 1.3^5 + 1.3^6 + \dots$

which is a GP: $a = 1.3^3$, $r = 1.3$, $n = 28$ (again!)

- $\sum_{r=3}^{30} 1.3^r = \frac{1.3^3(1 - 1.3^{28})}{1 - 1.3} = 11345.99\dots$

- Hence $\sum_{r=3}^{30} 3r - 5 + 1.3^r = 1246 + 11345.99\dots = 12591.99\dots = 12600$ (3 sf)

35 mins
Time Yourself

Series

Have you improved?

1 The seventh term of an arithmetic series is twice the eleventh term, and the fifth term of the series is 22.5.
a) Find the first term and the common difference of the series.
The sum of the first n terms of this series is 189.
b) Calculate the possible values of n.

2 a) Find the sum of all the numbers from 1 to 750 inclusive which are multiples of seven.
b) Hence find the sum of all the numbers from 1 to 750 inclusive that are not multiples of seven.

3 Jack invests £1200 at the beginning of *each* year into a ten-year savings plan. His financial advisor, Rhoda, tells him that at the end of each year, interest guaranteed at a fixed rate of 7% will be credited to his savings plan.
a) Calculate the value of Jack's investment at the end of:
i) the first year ii) the third year.
b) Write down a series that represents the total amount of Jack's investment at the end of the tenth year.
c) Name the type of series that you have written down.
d) Using the appropriate series formula, calculate to the nearest pound the value of Jack's investment when it matures at the end of the tenth year.

1a) Write down two equations in 'a' and 'd'. Solve them simultaneously

1b) Use S_n to find a quadratic. Solve it!

2a) Arithmetic Series

Common difference 7

2b) Work out sum of all numbers from 1 to 750 Then use part (a)

3) What multiple does £1200 increase by at the end of the year?

3a) Question is concerned with year-end value

3b) There should be a '+' between every term

3c) Constant ratio

3d) Use the S_n formula of the series you have chosen

Equations of Motion

How much do you know?

1 a) A car travelling from rest accelerates uniformly to a velocity of 18 ms^{-1} in 5 seconds. Calculate:

i) the acceleration of the car
ii) the distance covered in 5 seconds.

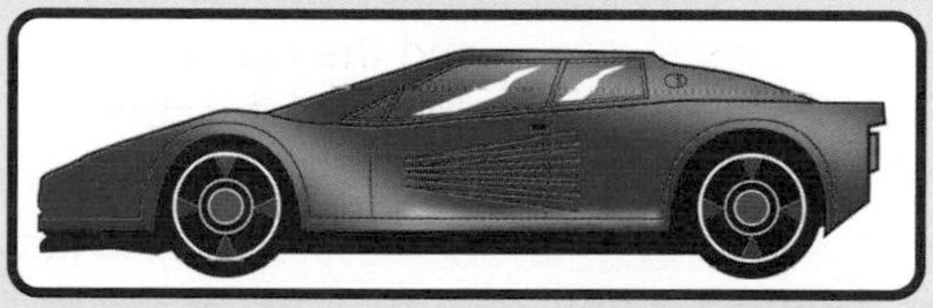

b) A particle P moves along a straight line with constant acceleration. Four seconds after passing through O, the particle passes through the point A, and twelve seconds after passing O it passes through the point B. Given that OA = 20 m and OB = 84 m, calculate:

i) the acceleration of the particle
ii) the velocity of the particle at the point O.

2 A stone is dropped from the top of a multi-storey car park, hitting the ground with velocity 18 ms^{-1}. Find, to 3 significant figures, the height of the multi-storey car park.

3 A bus travels between two check-points A and B, which are 330 metres apart. The bus accelerates uniformly at rest from the point A, until it reaches a velocity of 15 ms^{-1}. After travelling at a constant velocity of 15 ms^{-1} for 20 seconds, the bus passes through check-point B.

a) Sketch a velocity–time graph to illustrate the journey between A and B.
b) Find the time taken by the bus to travel between the two check-points.
c) Find the acceleration of the bus from check-point A.

Answers

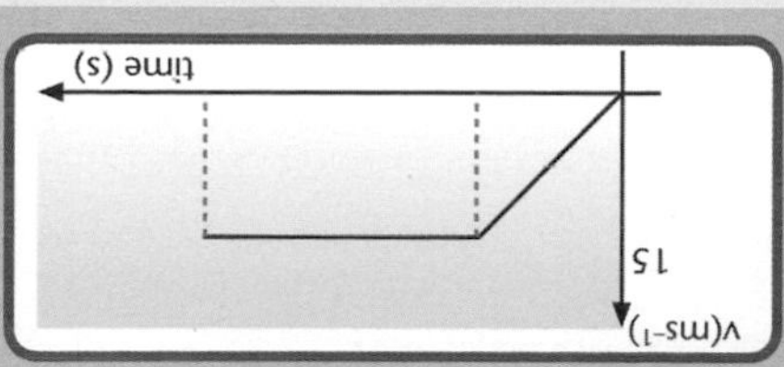

1a) (i) 3.6 ms^{-2} (ii) 45 m
b) (i) 0.5 ms^{-2} (ii) 4 ms^{-1}
2 16.5 m
3a) see diagram
b) 24 s c) 3.75 ms^{-2}

If you got them all right, skip to page 50

Spend no more than 45 mins on this topic

Equations of Motion

Learn the key facts

1 Motion in a Straight Line

You need to learn the following formulae when dealing with motion involving a constant (or uniform) acceleration:

In this chapter the acceleration (assumed constant) due to gravity is: $g = 9.8\ ms^{-2}$

- $s = \left(\dfrac{u+v}{2}\right).t$
- $v = u + at$
- $s = ut + \frac{1}{2}at^2$
- $v^2 = u^2 + 2as$

Where:
s is the displacement
a is the *constant* acceleration
u is the initial velocity
v is the final velocity
t is the time interval

It is useful to write down which letters you know, and then which letters you need to find. This helps you to choose the correct formula to use.

Example 1

A car travelling along a straight horizontal road accelerates from $10\,ms^{-1}$ to $26\,ms^{-1}$ in 8 seconds. Assuming the car is accelerating uniformly, calculate:
a) the acceleration of the car
b) the distance covered in these 8 seconds.

Solution

a) Given: $u = 10\ ms^{-1}$, $v = 26\ ms^{-1}$, $t = 8$ s. We want: a
- Formula: $v = u + at \Rightarrow 26 = 10 + 8a$
- Solve: $16 = 8a \Rightarrow a = 2\ ms^{-2}$

b) Formula: $s = ut + \frac{1}{2}at^2 \Rightarrow s = 10(8) + \frac{1}{2}(2)(8^2) = 144$ m

Example 2

A particle P moves along a straight line with constant acceleration. Two seconds after passing through the point O, the particle passes through the point A, and eight seconds later it passes through the point B. Given that OA = 17 m and
AB = 128 m, find:
a) the acceleration of the particle
b) the velocity of the particle at the point O.

Solution

a) When $t = 2$, $s = OA = 17$ m
- When $t = 10$, $s = OB = OA + AB = 17 + 128 = 145$ m

DAY 1 2 3 4 5 6 7

- We know the values of s and t; we want to find u and a $\Rightarrow$ Use $s = ut + \frac{1}{2}at^2$
- So: $17 = u(2) + \frac{1}{2}a(2^2) \Rightarrow 17 = 2u + 2a$ **(1)**
- And: $145 = u(10) + \frac{1}{2}a(10)^2 \Rightarrow 145 = 10u + 50a$ **(2)**

Solving simultaneously: **(2)** $- 5 \times$ **(1)** $\Rightarrow 60 = 40a \Rightarrow a = 1.5\ \text{ms}^{-2}$
Substitute back into **(1)**: $17 = 2u + 3 \Rightarrow 14 = 2u \Rightarrow u = 7\ \text{ms}^{-1}$

When a particle is decelerating, the value of the acceleration a is negative.

Example 3

A car travelling with constant velocity of $32\,\text{ms}^{-1}$ comes to rest, after applying the brakes. Given that the brakes produced a constant retardation of $5\,\text{ms}^{-2}$, calculate the distance covered by the car in coming to rest.

Solution

- Given: $u = 32\,\text{ms}^{-1}$, $v = 0\,\text{ms}^{-1}$ (at rest), $a = -5\,\text{ms}^{-2}$ (retardation). Want: s
- Formula: $v^2 = u^2 + 2as \Rightarrow 0^2 = (32)^2 + (2 \times -5 \times s) \Rightarrow 0 = 1024 - 10s$
- Solve: $1024 = 10s \Rightarrow s = 102.4$ metres.

Vertical Motion Under Gravity

2 When a particle (or object) is thrown vertically up, or dropped (to fall vertically downwards), then the particle moves with *constant* acceleration, g, due to gravity. Acceleration acts downwards; and when you solve such problems you should indicate which direction is positive.

Example 4

A particle is projected vertically upwards with velocity $21\,\text{ms}^{-1}$ from a point 5 metres vertically above a horizontal ground. Find, to one decimal place:

a) the greatest height of the particle above the ground
b) the times at which the particle is 20 m above the ground
c) the time taken for the particle to hit the ground.

Solution

a) Define ↑ as +ve because initial velocity is in an up direction. Then:

- Given: $u = 21$, $v = 0$ (at highest point) $a = -9.8$ (downwards). Want: $s = h$
- Formula: $v^2 = u^2 + 2as \Rightarrow 0^2 = 21^2 + (2 \times -9.8 \times h) \Rightarrow 0 = 441 - 19.6h$
- Solve: $19.6h = 441 \Rightarrow h = 22.5$ m

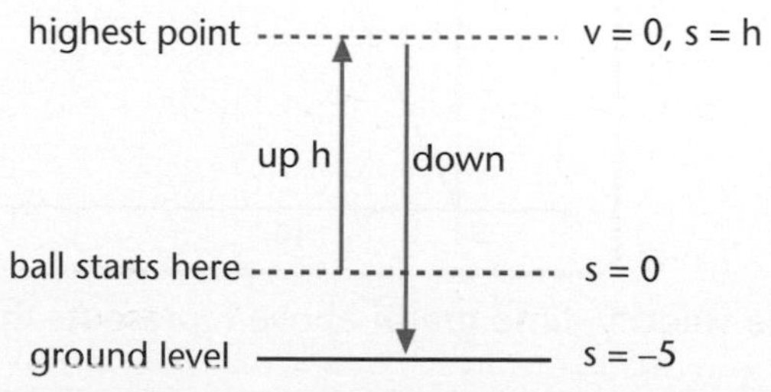

- Solve: $19.6h = 441 \Rightarrow h = 22.5$ m
- Hence height above ground = 5 + 22.5 = 27.5 metres.

b) 20 metres above the ground $\Rightarrow s = 20 - 5 = 15$ m

- Given: $u = 21, s = 15, a = -9.8$. Want: t
- Formula: $s = ut + \frac{1}{2}at^2 \Rightarrow 15 = 21t + \frac{1}{2}(-9.8)t^2 \Rightarrow 15 = 21t - 4.9t^2$
- Solve: $4.9t^2 - 21t + 15 = 0 \Rightarrow t = \dfrac{21 \pm \sqrt{441 - (4 \times 4.9 \times 15)}}{2 \times 4.9}$
- Hence: $t = 0.9$ s and $t = 3.4$ s

c) Ball hits ground $\Rightarrow$ the vertical displacement, $s = -5$ m

- Given: $u = 21, s = -5, a = -9.8$. Want: t
- Formula: $s = ut + \frac{1}{2}at^2 \Rightarrow -5 = 21t + \frac{1}{2}(-9.8)t^2 \Rightarrow -5 = 21t - 4.9t^2$
- Solve: $4.9t^2 - 21t - 5 = 0 \Rightarrow t = \dfrac{21 \pm \sqrt{441 - (4 \times 4.9 \times -5)}}{2 \times 4.9}$
- Hence: $t = -0.2$ s (reject) and $t = 4.5$ s (accept!)

DAY 4

3 Velocity–Time Graphs

A velocity–time graph has velocity on the y-axis and time on the x-axis.
For a velocity-time graph:

- Area between graph and x-axis represents total distance travelled
- The gradient represents the acceleration
- Positive gradient $\Rightarrow$ acceleration
- Negative gradient $\Rightarrow$ deceleration
- A horizontal line means that the velocity is constant $\Rightarrow$ no acceleration.

Example 5

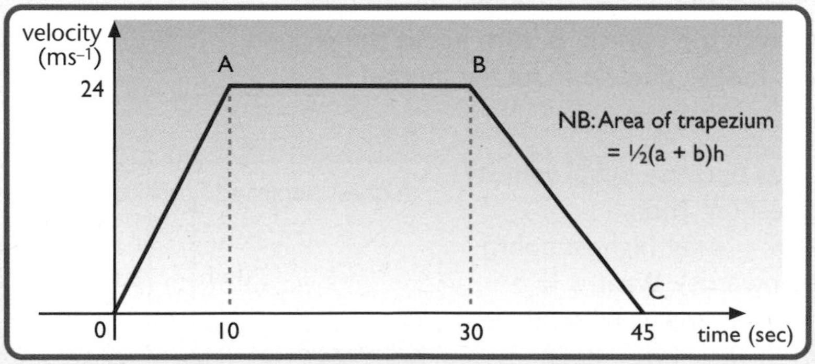

Area of trapezium $= \frac{1}{2}(a+b)h$

The velocity–time graph above represents the short journey of a car.

a) Find the acceleration of the car between:
   i) 0 and 10 seconds   ii) 10 and 30 seconds   iii) 30 and 45 seconds.
b) Find the total distance travelled.
c) Sketch the corresponding acceleration–time graph.

Solution

a) i) accel = gradient OA = $(24 - 0)/(10 - 0) = 2.4\ \text{ms}^{-2}$
   ii) accel = gradient AB = $0\ \text{ms}^{-2}$ (because it's a horizontal line!)
   iii) accel = gradient BC = $-\frac{24}{15} = -1.6\ \text{ms}^{-2}$

b) Distance = area of trapezium OABC = $\frac{1}{2}(45+20)\times 24 = 780\ \text{m}$

c) Using our answers to part a) we can construct the acceleration–time graph, to the right.

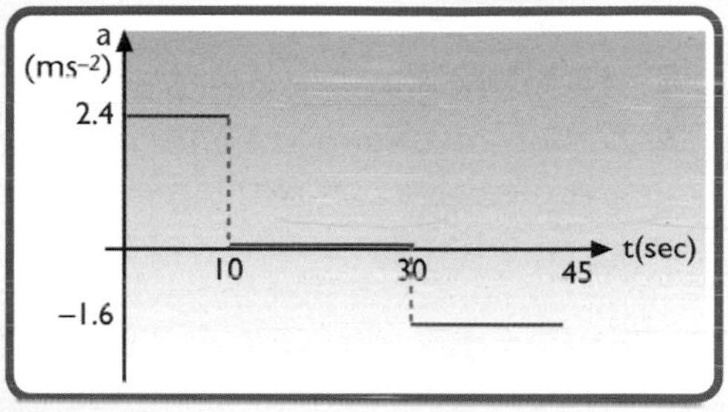

Example 6

A tram stops at two points A and B that are 744 metres apart. It accelerates uniformly from A for 96 metres reaching a velocity of $24\ \text{ms}^{-1}$. The velocity is maintained for T seconds. Then the tram uniformly decelerates for 10 seconds to arrive at the point B.
a) Sketch a velocity–time graph to illustrate the journey from A to B.
b) Using your graph find the value of T.
c) Hence find the time taken for the tram to travel from A to B.

Solution

a) See opposite.
b) Area of triangle WBY = $\frac{1}{2} \times 24 \times 10 = 120\ \text{m}$
   (≡ deceleration distance)
   - From sketch:
     Total distance = $96 + 24T + 120 = 744$
   - Solve: $24T = 744 - 216 = 528 \Rightarrow$
     $T = 22$
c) From sketch: Time taken = $8 + 22 + 10$
   = 40 seconds.

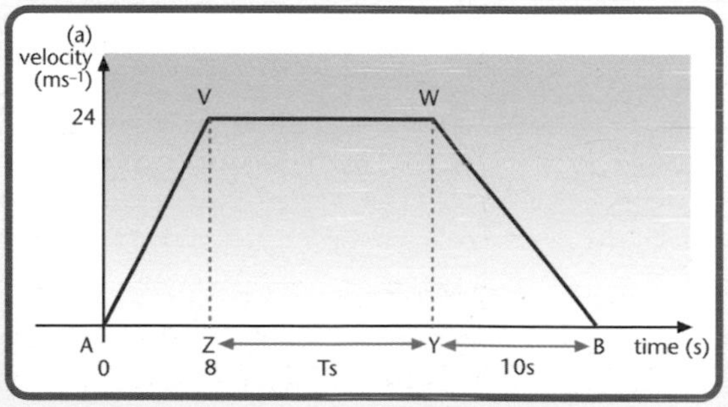

Equations of Motion

Have you improved?

1

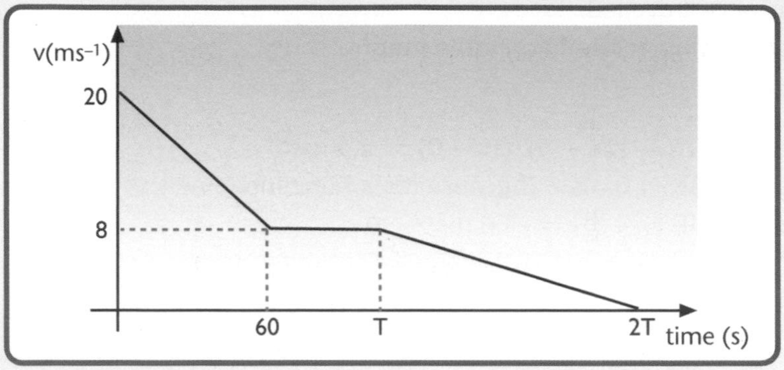

The graph above is a sketch of the velocity–time graph modelling the motion of a train, just after it passes through a signal, when $t = 0$, to when it comes to a halt at the railway station when $t = 2T$.

a) Find the deceleration of the train 30 seconds after it passes through the signal. Given that the average speed from $t = 0$ to $t = 2T$ is $7.8\,\text{ms}^{-1}$:
b) Find the value of T.
c) Deduce the time it takes the train from passing through the signal to stopping at the railway station.

1a) Gradient on the graph at that time.

1b) Area under graph?

1c) Use part b) and look at the sketch for help

2 A particle is projected vertically upwards with speed $24.5\,\text{ms}^{-1}$ from a point 2 metres above horizontal ground.
Find, to 2 decimal places, if necessary:
a) the greatest height of the particle above the ground
b) the length of time for which the particle is at least 21.6 metres above the ground
c) the time taken for the particle to hit the ground.

2a) Occurs when $v = 0$. Remember to add on 2

2b) 21.6 above ground $\equiv s = 21.6 - 2 = 19.6$

2c) Find out the time when $s = -2$

Equilibrium, Friction and Newton's Laws

40 mins
Time Yourself

How much do you know?

1 a) Find the horizontal and vertical components of this force.
b) Find the components of this force along and perpendicular to the plane.

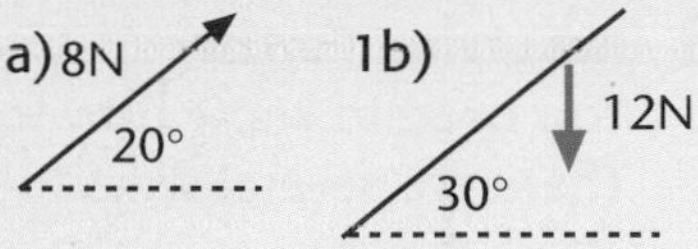

2 Find the resultant of each of the sets of forces in diagrams 2a) and 2b).

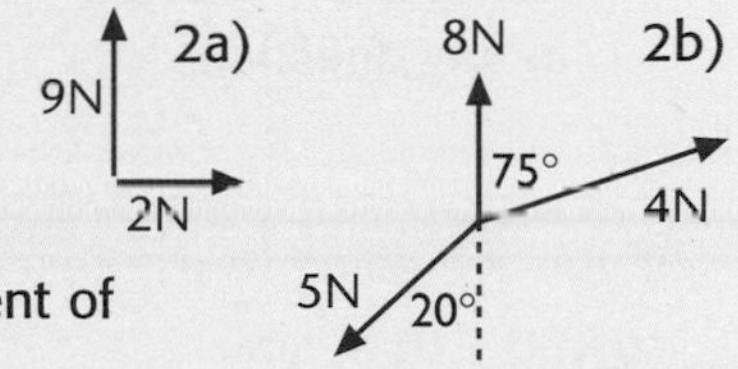

3 a) Explain what limiting friction means.
b) With the aid of an equation, explain what the coefficient of friction is.

4 A block of mass 6 kg is resting in limiting equilibrium on a rough plane inclined at 25° to the horizontal. Find the coefficient of friction between the block and the plane.

5 a) A motorcycle of mass 400 kg is accelerating down a hill inclined at 10° to the horizontal. The frictional resistances to motion are constant and of magnitude 400 N. The motorcycle's engine exerts a driving force of 500 N. Find the acceleration of the motorcycle.
b) A railway engine of mass 2000 kg is towing a carriage of mass 600 kg along level rails. Both engine and carriage have frictional resistances to motion of 1200 N. The driving force of the engine is 3000 N. Find the acceleration of the engine and carriage and the tension in the towbar.

6 Particles of mass 6 kg and 4 kg are attached to the ends of a light inextensible string, which is passed over a smooth pulley that is suspended above the ground. Find the acceleration of the particles and the tension in the string when the system is released.

Answers

1a) H: 7.52 N, V: 2.74 N   b) along: 6 N perp. 10.4 N
2a) 9.22 N at 77° above horizontal   b) 4.84 N at 64° above horizontal
3a) The maximum value friction can take; occurs when object is about to slip.
b) When friction is limiting, $F = \mu R$, where R is the normal reaction and μ is the coefficient of friction.
4 0.466   **5a)** 1.95 ms^{-2} b) $a = 0.231$ ms^{-2}, $T = 1340$ N (3 SF)
6 $a = 1.96$ ms^{-2}, $T = 47.04$ N

If you got them all right, skip to page 58

Spend no more than 60 mins on this topic

Equilibrium, Friction and Newton's Laws

Learn the key facts

1 Resolving Forces

You need to know how to resolve forces into components in any direction. There are several ways to do this. Here is one!

Component of force in any direction	=	size of of force	×	cosine of angle between force and the direction you're resolving in.

If you are going to use this method, you also need to remember the rule:

$\cos(90 - \theta) = \sin\theta$

The diagrams illustrate how to use this:

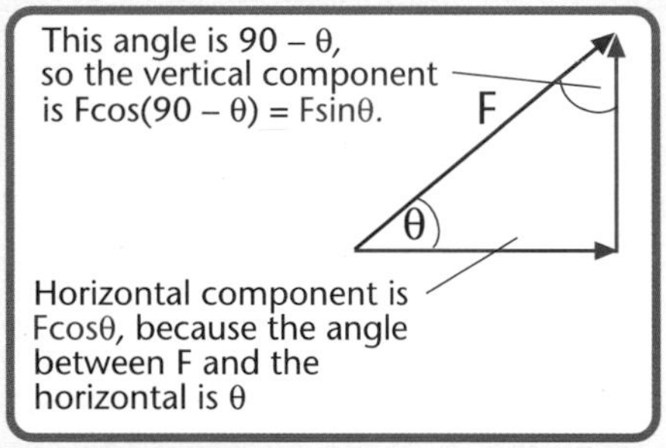
This angle is 90 – θ, so the vertical component is Fcos(90 – θ) = Fsinθ.
F
θ
Horizontal component is Fcosθ, because the angle between F and the horizontal is θ

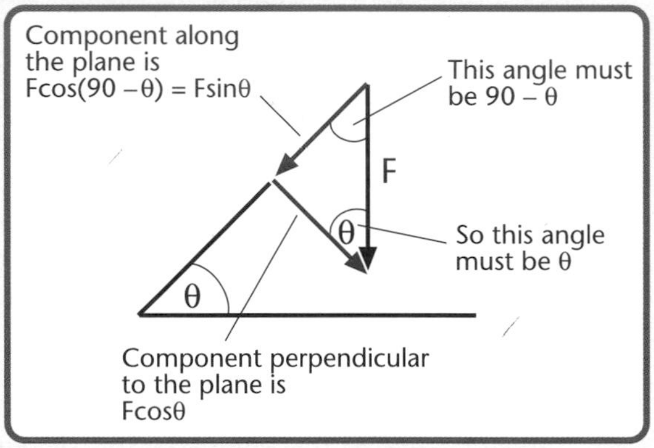
Component along the plane is Fcos(90 –θ) = Fsinθ
This angle must be 90 – θ
F
θ
So this angle must be θ
θ
Component perpendicular to the plane is Fcosθ

2 Resultant Forces

The resultant force on something is the combined effect of all the forces acting on it. To find the resultant of two perpendicular forces, we use Pythagoras and trig.

Example 1

Find the resultant of a force of 80 N acting horizontally and a force of 60 N acting vertically.

Solution

To find the resultant force, we must give its length (or magnitude) and direction.

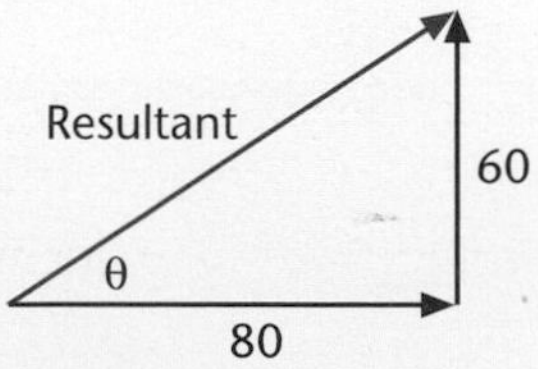
Resultant
60
θ
80

By Pythagoras, length = $\sqrt{60^2 + 80^2} = 100$ N

Direction = $\tan^{-1}\left(\frac{60}{80}\right) = 37^{\circ}$ above horizontal

DAY 5

To find the resultant of more than two forces, or of two forces that are not at right angles, you follow the following steps:

1) Choose two convenient perpendicular directions (such as horizontal and vertical, or along and perpendicular to a plane) and resolve each force in these directions. Make sure you are careful with minus signs – decide in advance which direction is positive and which is negative.
2) Add up all the components in each direction.
3) Find the resultant of the two perpendicular forces you get.

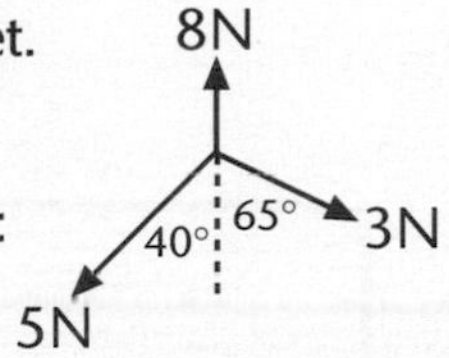

Example 2

Find the resultant of the set of forces shown in the diagram:

Solution

We will resolve each force horizontally and vertically, and we will choose to take upwards and to the right as positive.

Force	Horizontal component	Vertical component
8N	0	8
3N	$3\sin65^0$	$-3\cos65^0$
5N	$-5\sin40^0$	$-5\cos40^0$
Total	−0.4950	2.9019

So to find the resultant:

Magnitude of resultant $= \sqrt{2.9019^2 + 0.4950^2}$

$= 2.944$ N

at $\tan^{-1}\left(\frac{2.9019}{0.4950}\right) = 80.3°$ above the negative x-axis

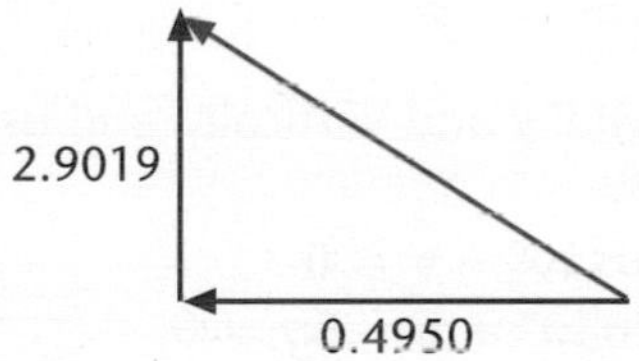

Drawn in this direction because of the minus

3 Friction

Friction occurs whenever we have a 'rough' surface.

- It always acts to oppose motion, or any tendency to motion.
- If there is no tendency to motion there is no friction.
- Friction will increase as necessary to prevent motion, but it can only increase up to its limiting value. Its limiting value is given by $F = \mu R$, where μ is the coefficient of friction and R is the normal reaction.
- When a body is actually moving, friction takes its limiting value so $F = \mu R$.

4 Equilibrium

A body is in equilibrium if there is no resultant force on it. This happens if it is not moving or is moving at constant velocity. (This is Newton's first law.)
So if something is in equilibrium the forces on it balance. We use this by resolving all the forces in two perpendicular directions and saying that the components must add to zero in each direction.
It's ALWAYS worth resolving in both directions, even if you're not convinced you'll need both.

Example 3

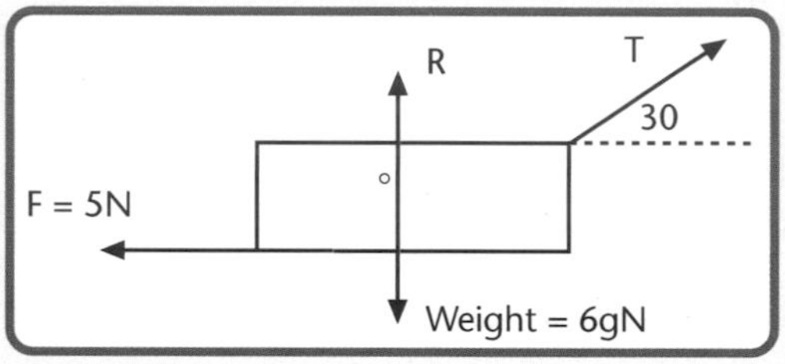

Remember: weight is mass × g

A box of mass 6 kg is being towed at a constant speed along a rough horizontal plane by a rope inclined at 30° to the horizontal. The frictional force on the box is constant and of magnitude 5 N. Find:

a) the tension in the rope
b) the coefficient of friction between the box and the plane.

Solution

Resolving horizontally and vertically, and using the fact the box is in equilibrium:

Horizontally: $T\cos 30° - 5 = 0$ **(1)**
Vertically: $T\sin 30° + R - 6g = 0$ **(2)**

a) We want T, so use equation **(1)**. We get $T = \dfrac{5}{\cos 30} = \dfrac{5}{\frac{\sqrt{3}}{2}} = \dfrac{10}{\sqrt{3}} = 5.77\,\text{N}$

b) Since we are asked for the coefficient of friction, we must use $F = \mu R$, since it is moving. As we need to find R, use equation **(2)**:

We get $R = 6g - T\sin 30 = 55.91$ N. So $\mu = \dfrac{F}{R} = \dfrac{5}{55.91} = 0.0894$ (3 SF)

Example 4

A particle of mass m is held in position on a rough plane inclined at 30° to the horizontal by a horizontal force of magnitude kmg. The particle is on the point of slipping down the plane. The coefficient of friction between the particle and the plane is 0.2. Find the value of k.

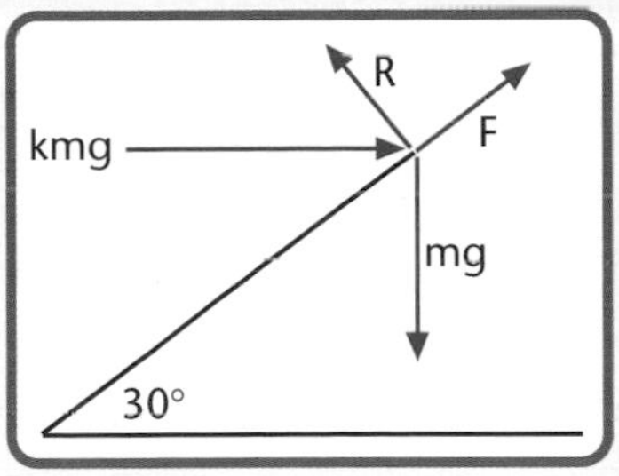

'On the point of slipping' means friction is limiting

Always resolve along and perpendicular to the plane

Solution

Resolving, and using the fact the particle is in equilibrium:
Along plane: $kmg\cos30° + F - mg\sin30° = 0$ **(1)**
Perp. to plane: $R - kmg\sin30° - mg\cos30° = 0$ **(2)**
We also know, because particle is in limiting equilibrium, that $F = 0.2R$
Rearranging **(1)** and **(2)** and substituting into $F = 0.2R$, we get:
$mg\sin30° - kmg\cos30° = 0.2(kmg\sin30° + mg\cos30°)$

Cancelling mg, and rearranging, we get $k = \dfrac{\sin 30 - 0.2\cos 30}{0.2\sin 30 + \cos 30} = 0.338$ (3 SF)

5 Newton's Second Law

Newton's second law says: $F = ma$ where F is the resultant force in the direction the body is moving in, and a is the acceleration.

NB: The resultant force perpendicular to the direction of motion will be zero, because there is no acceleration in that direction. You often need to use this!

Some problems involving use of Newton's second law require you to do calculations after you have found the acceleration – look at the chapter on 'Equations of Motion' to remind you how to do these!

Example 5

A car of mass 1000 kg is ascending a road inclined at $\sin^{-1}(0.1)$ to the horizontal. The coefficient of friction between the car and the road is 0.6. Find the driving force of the car when it is accelerating at $0.2\,\text{ms}^{-2}$.

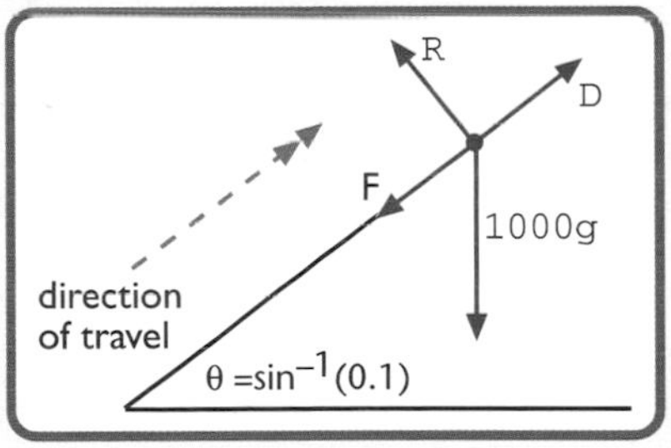

DAY 5

Solution

There's no resultant force perpendicular to the plane, so resolving in that direction:

$R - 1000g\cos\theta = 0$ **(1)**

Resolving in the direction of motion:

Resultant force $= D - F - 1000g\sin\theta$

Using $F = ma$, we get: $D - F - 1000g\sin\theta = 1000 \times 0.2$ **(2)**

Since the car is moving, we have $F = \mu R$, so from **(1)**, $F = 0.6 \times 1000g\cos\theta$ **(3)**

Substituting **(3)** into **(2)** and rearranging:

$D = 0.6 \times 1000g\cos\theta + 1000g\sin\theta + 1000 \times 0.2 = 7030$ N (3 SF)

Example 6

A car of mass 900 kg is towing a trailer of mass 200 kg along a level road. The frictional resistances to motion, which are constant, are 500 N for the car and 300 N for the trailer. The tension in the tow bar connecting the car and trailer is 450 N. Find:

a) the acceleration of the trailer

b) the driving force of the car.

Solution

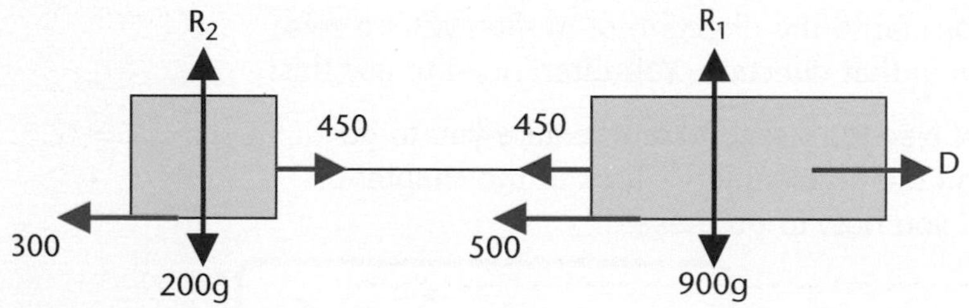

a) For trailer, using $F = ma$: $450 - 300 = 200a \Rightarrow a = 0.75\,\text{ms}^{-2}$

b) For car, using $F = ma$: $D - 450 - 500 = 900 \times 0.75 \Rightarrow D = 1625\,\text{N}$

6 Pulleys

If a string passes over a smooth pulley, then:

- The tension is the same on each side of the pulley.
- The particles on each side have the same speed and acceleration.

The strategy for pulley questions is to resolve in the direction of motion for each particle, use $F = ma$, and then use simultanous equations to find unknown forces or accelerations.

You can consider connected objects separately

Connected objects must have the same speed and acceleration

Example 7

Particles P and Q have masses 2 kg and 3 kg respectively. They are fastened to either end of a light inextensible string which is passed over a smooth pulley that is suspended above the ground. The particles are initially held at the same level, then released. Find the acceleration of the particles and the tension in the string in terms of g.

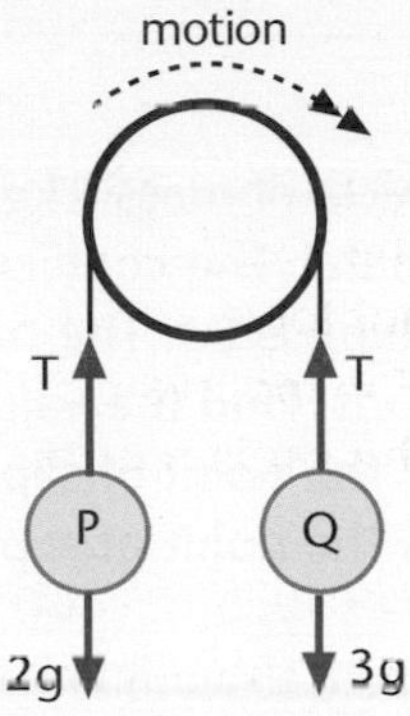

Solution

For P: $T - 2g = 2a$ **(1)**

For Q: $3g - T = 3a$ **(2)**

(1) + **(2)**: $3g - 2g = 5a \Rightarrow a = \frac{1}{5}g$

Substituting into **(1)**: $T = \frac{12}{5}g$

Example 8

A particle of mass $4m$ is held on a smooth wedge inclined at 30° to the horizontal. It is attached to a light, inextensible string which passes over a smooth pulley at the top of the wedge. The other end of the string is attached to a particle of mass $6m$, which hangs freely. Find the acceleration of the particles when they are released from rest.

Solution

For $6m$ particle: $6mg - T = 6ma$ **(1)**

For $4m$ particle: $T - 4mg\sin 30° = 4ma$ **(2)**

(1) + **(2)**: $6mg - 4mg\sin 30° = 10ma \Rightarrow a = 0.4g = 3.92\,\text{ms}^{-2}$

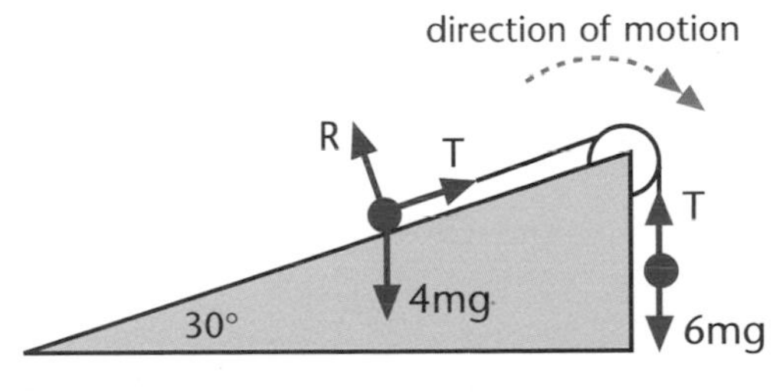

DAY 5

Equilibrium, Friction and Newton's Laws

Have you improved?

1 A particle of mass 2 kg rests on a rough plane inclined at $\tan^{-1}(\frac{3}{4})$ to the horizontal. The coefficient of friction between the particle and the plane is 0.1. The particle is held in equilibrium by a horizontal force of P newtons. Find the value of P if the particle is:

a) on the point of slipping down the plane

b) on the point of slipping up the plane.

2 A particle rests in limiting equilibrium on a plane inclined at angle θ to the horizontal. The coefficient of friction between the particle and the plane is μ. Prove that $\tan\theta = \mu$.

3 Particles A and B are of masses 0.4 kg and 0.6 kg respectively. They are connected to either end of a light inextensible string of length 1.2 metres. Particle A is held on a smooth horizontal table 80 cm from the edge. The string passes over a smooth pulley on the edge of the table, and particle B hangs freely.

a) Find the acceleration of the particles when they are released from rest.

b) Find the force exerted on the pulley.

Given that the table is 80 cm high:

c) Find the velocity with which B strikes the floor.

1a) Friction is going up the plane. Resolve along and perpendicular to the plane. Draw a triangle to find $\sin\theta$ and $\cos\theta$. Use $F = \mu R$

1b) Friction is going down the plane this time!

2) Resolve along and perpendicular to the plane and use $F = \mu R$. Remember $\tan = \sin \div \cos$

3a) Resolve in the direction of motion of each particle and do simultaneous equations

3b) The force on the pulley is due to the horizontal and vertical tensions

3c) This is an equations of motion question!

Vectors in Mechanics

20 mins
Time Yourself

How much do you know?

1 $\mathbf{a} = 4\mathbf{i} - 6\mathbf{j}$ $\quad \mathbf{b} = -2\mathbf{i} + 8\mathbf{j}$

a) Find: i) $\mathbf{a} + \mathbf{b}$ ii) $\mathbf{a} - \mathbf{b}$ iii) $4\mathbf{b}$ iv) $3\mathbf{a} - 5\mathbf{b}$

b) Find the magnitude of **a** and a unit vector in the direction of **a**

c) $\mathbf{c} = k\mathbf{i} - (4 - k)\mathbf{j}$. If **a** and **c** are parallel, find k.

d) $\mathbf{d} = (2t - 4)\mathbf{i} + 6t\mathbf{j}$. Show that **a** and **d** are never equal.

2 a) Find the resultant of these forces: $\mathbf{F}_1 = 5\mathbf{i} - 6\mathbf{j}$, $\mathbf{F}_2 = 7\mathbf{i} - 5\mathbf{j}$, and $\mathbf{F}_3 = -4\mathbf{i} + 10\mathbf{j}$

b) These forces act on a particle of mass 2 kg. Find the magnitude of the acceleration produced.

3 a) A particle is initially at the origin, moving with velocity $6\mathbf{i} - 5\mathbf{j}$ and its acceleration is $-2\mathbf{i} + \mathbf{j}$. Find its speed and distance from the origin after two seconds.

b) In this question, **i** and **j** are unit vectors East and North respectively. At noon, ship A is at point O on the coast and ship B is at the point with position vector $6\mathbf{i} + \mathbf{j}$ km moving with velocity $-\mathbf{i} + 2\mathbf{j}$ kmh^{-1}. A sets off at constant velocity **v** and intercepts B at 4pm. Find **v**.

Answers

1a) i) $2\mathbf{i} + 2\mathbf{j}$ ii) $6\mathbf{i} - 14\mathbf{j}$ iii) $-8\mathbf{i} + 32\mathbf{j}$ iv) $22\mathbf{i} - 58\mathbf{j}$

b) $\sqrt{52}$ ($= 2\sqrt{13}$), $\frac{1}{\sqrt{13}}(2\mathbf{i} - 3\mathbf{j})$ c) 1.6

d) **i** parts equal when $t = 4$. **j** parts equal when $t = -1$. t cannot take both values at once ⇒ **a** and **d** never equal.

2a) $8\mathbf{i} - \mathbf{j}$ b) 4.03 ms^{-2}

3a) speed $= \sqrt{13}$ ms^{-1}, distance $= 8\sqrt{2}$ m b) $0.5\mathbf{i} + 2.25\mathbf{j}$ ms^{-1}

If you got them all right, skip to page 65

Spend no more than 45 mins on this topic

Vectors in Mechanics

Learn the key facts

1 Basic Vector Ideas and Techniques

The vectors you will come across in mechanics will be in terms of $\mathbf{i}$ and $\mathbf{j}$, which are vectors of length one unit going across and up respectively. They are very like coordinates – for example, if an object is at the point (2, 3), we could say its position vector is $2\mathbf{i} + 3\mathbf{j}$. Similarly, an object with position vector $6\mathbf{i} - 5\mathbf{j}$ is at the point (6, –5).

In mechanics, you need to be able to:

- add and subtract vectors
- multiply and divide vectors by a number
- find the length (or magnitude) of a vector
- find a unit vector in the direction of a vector you are given
- understand how to tell whether two vectors are parallel
- understand how to tell whether two vectors are equal.

To add/subtract vectors, you add/subtract the $\mathbf{i}$ parts and add/subtract the $\mathbf{j}$ parts. To multiply or divide a vector by a number, you multiply/divide each part by it.

Example 1

$\mathbf{a} = 6\mathbf{i} + 4\mathbf{j}$ $\quad \mathbf{b} = -2\mathbf{i} - 8\mathbf{j}$

Find: i) $\mathbf{a} + \mathbf{b}$ ii) $\mathbf{a} - \mathbf{b}$ iii) $3\mathbf{a}$ iv) $2\mathbf{a} + 4\mathbf{b}$

Solution

i) $\mathbf{a} + \mathbf{b} = 6\mathbf{i} + 4\mathbf{j} + (-2\mathbf{i} - 8\mathbf{j}) = (6 + -2)\mathbf{i} + (4 + -8)\mathbf{j} = 4\mathbf{i} - 4\mathbf{j}$

ii) $\mathbf{a} - \mathbf{b} = 6\mathbf{i} + 4\mathbf{j} - (-2\mathbf{i} - 8\mathbf{j}) = 6\mathbf{i} + 4\mathbf{j} + 2\mathbf{i} + 8\mathbf{j} = 8\mathbf{i} + 12\mathbf{j}$

iii) $3\mathbf{a} = 3(6\mathbf{i} + 4\mathbf{j}) = 18\mathbf{i} + 12\mathbf{j}$

iv) $2\mathbf{a} + 4\mathbf{b} = 2(6\mathbf{i} + 4\mathbf{j}) + 4(-2\mathbf{i} - 8\mathbf{j}) = 12\mathbf{i} + 8\mathbf{j} - 8\mathbf{i} - 32\mathbf{j} = 4\mathbf{i} - 24\mathbf{j}$

To find the magnitude of a vector, you are effectively using Pythagoras:

$$\text{magnitude} = \sqrt{(\text{number in front of i})^2 + (\text{number in front of j})^2}$$

You NEVER square the i or j – just the numbers!

To find a unit vector (that's a vector of length 1) in a particular direction, you find any vector in that direction, find its magnitude, and divide the vector by it.

$$\text{unit vector} = \frac{\text{vector}}{\text{its magnitude}}$$

DAY 5

Example 2

a) Find the magnitude of $5\mathbf{i} - 12\mathbf{j}$
b) Find a unit vector in the direction of $5\mathbf{i} - 12\mathbf{j}$

Solution

a) magnitude $= \sqrt{5^2 + (-12)^2} = \sqrt{169} = 13$

b) unit vector $= \dfrac{5\mathbf{i} - 12\mathbf{j}}{13} = \dfrac{5}{13}\mathbf{i} - \dfrac{12}{13}\mathbf{j}$

If you need to find the magnitude of a sum or difference of vectors, always combine first then find the magnitude!

Two vectors are parallel if their **i** and **j** parts are in the same ratio. An easy way to tell if this is true is to see whether

LEARN!

$$\frac{\text{i part of first vector}}{\text{i part of second vector}} = \frac{\text{j part of first vector}}{\text{j part of second vector}}$$

Special case: Any vector is parallel to **i** if its **j** part is zero. Similarly, any vector is parallel to **j** if its **i** part is zero.

Two vectors are equal if their **i** parts are equal and their **j** parts are equal.

LEARN!

Example 3

a) **a** and **b** are parallel. $\mathbf{a} = k\mathbf{i} + 2\mathbf{j}$ $\quad \mathbf{b} = 9\mathbf{i} + (3k - 3)\mathbf{j}$.
Find the value(s) of k.
b) $\mathbf{c} = (2t - 1)\mathbf{i} + (4 - t)\mathbf{j}$ $\quad \mathbf{d} = 3\mathbf{i} + 2t\mathbf{j}$
Show that **c** and **d** are never equal.

Solution

a) We use $\dfrac{k}{9} = \dfrac{2}{3k - 3}$ since the vectors are parallel.

Cross multiplying gives: $k(3k - 3) = 18 \Rightarrow 3k^2 - 3k - 18 = 0$
Factorising gives us: $3(k - 3)(k + 2) = 0$, giving $k = 3$ or $k = -2$

b) Suppose **c** and **d** are equal. Then we have
$(2t - 1)\mathbf{i} + (4 - t)\mathbf{j} = 3\mathbf{i} + 2t\mathbf{j}$
i parts equal, so we must have: $2t - 1 = 3$ **(1)**
j parts equal, so we must have: $4 - t = 2t$ **(2)**
(1) $\Rightarrow t = 2$
(2) $\Rightarrow t = \dfrac{4}{3}$

To show things are never equal, start out assuming they are and see what happens!

But t can't have two values at once, so **c** and **d** are never equal.

1

2

3

4

DAY

5

6

7

2 Forces as Vectors

Forces may be given as vectors. This makes it easier to find the resultant force – you just add up the vectors. You use $F = ma$ in exactly the same way as well – the acceleration you get will be a vector.

Example 4

A particle is acted on by $\mathbf{F}_1 = 6\mathbf{i} + 5\mathbf{j}$, $\mathbf{F}_2 = -4\mathbf{i} + 3\mathbf{j}$ and $\mathbf{F}_3 = a\mathbf{i} + b\mathbf{j}$,

a) Given that the particle is in equilibrium, find a and b.

b) $\mathbf{F}_3$ is now removed and replaced by $\mathbf{F}_4 = 2\mathbf{i} - 6\mathbf{j}$. Given that all forces are in newtons, and that the particle's mass is 2 kg, find the magnitude of the particle's acceleration.

Solution

a) The particle is in equilibrium, so there is no resultant force, ⇒ forces sum to zero. $6\mathbf{i} + 5\mathbf{j} + -4\mathbf{i} + 3\mathbf{j} + a\mathbf{i} + b\mathbf{j} = 0 \Rightarrow a = -2$ and $b = -8$

b) Sum of forces = resultant force = $6\mathbf{i} + 5\mathbf{j} + -4\mathbf{i} + 3\mathbf{j} + 2\mathbf{i} - 6\mathbf{j} = 4\mathbf{i} + 2\mathbf{j}$
So using $\mathbf{F} = m\mathbf{a}$: $4\mathbf{i} + 2\mathbf{j} = 2\mathbf{a} \Rightarrow \mathbf{a} = 2\mathbf{i} + \mathbf{j}$

We need the magnitude: magnitude = $\sqrt{2^2 + 1^2} = \sqrt{5}$ ms^{-2}

3 The Equations of Motion for Vectors

Here are the vector equivalents of the equations of motion:

$\mathbf{v} = \mathbf{u} + \mathbf{a}t$

$\mathbf{s} = \mathbf{u}t + \frac{1}{2}\mathbf{a}t^2$

Here is another one that you may not have seen before but is very useful. **It is only true if velocity is constant!**

$$\mathbf{r} = \mathbf{r_0} + \mathbf{v}t$$

Where it is now – its position vector   Where it started off when t was zero

Displacement, velocity and acceleration are all vectors. Time isn't!

Here are some facts and methods that are useful in this sort of problem:

- Two things collide if they are at the same place at the same time – in other words, their position vectors are the same for a particular value of t.
- To find the distance between two objects whose position vectors you know, convert the position vectors to coordinates and use the distance formula for them.
- Distance from the origin is just the magnitude of the position vector.
- The speed is the magnitude of the velocity vector.
- If you know the direction of the velocity and you know the speed, you can find the velocity vector by finding a unit vector in the correct direction and multiplying it by the speed.

Example 5

A particle of mass 1.5 kg is acted on by a force $\mathbf{F} = 6\mathbf{i} - 9\mathbf{j}$. Initially the particle is at the origin and has velocity $4\mathbf{i} + \mathbf{j}$. Find: i) its speed; ii) its position vector when $t = 4$.

Solution

i) First we must find the acceleration: $\mathbf{F} = m\mathbf{a} \Rightarrow \mathbf{a} = \frac{1}{1.5}\mathbf{F} = 4\mathbf{i} - 6\mathbf{j}$

Using $\mathbf{v} = \mathbf{u} + \mathbf{a}t$: $\mathbf{v} = 4\mathbf{i} + \mathbf{j} + (4\mathbf{i} - 6\mathbf{j}) \times 4 = 4\mathbf{i} + \mathbf{j} + 16\mathbf{i} - 24\mathbf{j} = 20\mathbf{i} - 23\mathbf{j}$

But we need speed = magnitude of velocity = $\sqrt{20^2 + (-23)^2} = 30.5 \text{ ms}^{-1}$

ii) Using $\mathbf{s} = \mathbf{u}t + \frac{1}{2}\mathbf{a}t^2$: $\mathbf{s} = (4\mathbf{i} + \mathbf{j}) \times 4 + \frac{1}{2}(4\mathbf{i} - 6\mathbf{j}) \times 4^2 = 48\mathbf{i} - 44\mathbf{j}$

Example 6

Andrew and Jaspal are standing initially at the points P and Q which have position vectors $4\mathbf{i} + 5\mathbf{j}$ and $-6\mathbf{i} + \mathbf{j}$ respectively. Andrew starts to run with velocity $2\mathbf{i} - \mathbf{j}$. Jaspal starts to run with velocity $\mathbf{v}$. They meet after 10 seconds.

a) Find $\mathbf{v}$.
b) Find the distance between Jaspal and Andrew after 2 seconds.
c) Beth starts at the point R, which has position vector $3\mathbf{i} - 7\mathbf{j}$. If she starts to run with velocity $-\mathbf{i} + 3\mathbf{j}$ at the same time as Andrew and Jaspal, show she can never meet Andrew.
d) If, instead, she starts to run with speed 5 ms^{-1} in order to reach the point S with position vector $-7\mathbf{i} + 17\mathbf{j}$, find her velocity vector.

Solution

a) We will use $\mathbf{r} = \mathbf{r_0} + \mathbf{v}t$, putting in $t = 10$.

For Andrew: $\mathbf{r}_A = 4\mathbf{i} + 5\mathbf{j} + (2\mathbf{i} - \mathbf{j}) \times 10 = 24\mathbf{i} - 5\mathbf{j}$

For Jaspal: $\mathbf{r}_J = -6\mathbf{i} + \mathbf{j} + 10\mathbf{v}$

When they meet, we must have $\mathbf{r}_A = \mathbf{r}_J$

So $24\mathbf{i} - 5\mathbf{j} = -6\mathbf{i} + \mathbf{j} + 10\mathbf{v} \Rightarrow \mathbf{v} = 3\mathbf{i} - 0.6\mathbf{j}$

ALWAYS use this equation if there is no acceleration and you are given an initial position

1 2 3 4 DAY 5 6 7

b) When t = 2, $\mathbf{r}_A = 4\mathbf{i} + 5\mathbf{j} + (2\mathbf{i} - \mathbf{j}) \times 2 = 8\mathbf{i} + 3\mathbf{j}$,
$\mathbf{r}_J = -6\mathbf{i} + \mathbf{j} + (3\mathbf{i} - 0.6\mathbf{j}) \times 2 = -0.2\mathbf{j}$
So their 'coordinates' are (8, 3) and (0, −0.2)

So the distance between them is $\sqrt{(8-0)^2 + (3--0.2)^2} = 8.62\,\text{m}$

Combine i parts and j parts before you try to do anything!

c) At any time t, we have: $\mathbf{r}_A = 4\mathbf{i} + 5\mathbf{j} + (2\mathbf{i} - \mathbf{j})t = (4 + 2t)\mathbf{i} + (5 - t)\mathbf{j}$,
$\mathbf{r}_B = 3\mathbf{i} - 7\mathbf{j} + (-\mathbf{i} + 3\mathbf{j})t = (3 - t)\mathbf{i} + (-7 + 3t)\mathbf{j}$
Suppose Beth meets Andrew at some time t. Then we'd have $\mathbf{r}_A = \mathbf{r}_B$.
Look at **i** parts: $4 + 2t = 3 - t$ **(1)**
j parts: $5 - t = -7 + 3t$ **(2)**
(1) $\Rightarrow t = -\frac{1}{3}$, **(2)** $\Rightarrow t = 3$
These are not the same, so they can never meet.

d) Beth starts at $3\mathbf{i} - 7\mathbf{j}$ and she wants to reach $-7\mathbf{i} + 17\mathbf{j}$
To do this, she must go $-10\mathbf{i} + 24\mathbf{j}$
So her velocity must be in that direction.
So her velocity = speed × unit vector in correct direction

$$= 5 \times \frac{-10\mathbf{i} + 24\mathbf{j}}{\sqrt{(-10)^2 + 24^2}} = \frac{5}{26}(-10\mathbf{i} + 24\mathbf{j})$$

30 mins
Time Yourself

Have you improved?

1 At 8am, ship A is 8 km due north of point O on the coast, and ship B is $7\sqrt{2}$ km northwest of O. The ships set off with velocities **u** kmh^{-1} and **v** kmh^{-1} respectively. Taking **i** and **j** to be unit vectors east and north respectively:

a) Write down expressions for the positions of A and B relative to O when T hours have passed after 8am in terms of **u** and **v**.

Both ships are due to reach point P at 1 pm, where $\overrightarrow{OP} = -17\mathbf{i} + 17\mathbf{j}$

b) Find the speed of each ship.

Ship A meets with an accident at 9am and stops. Ship B immediately changes course to head for ship A.

c) Assuming the speeds found in b) are the maximum speeds for each ship, find the time at which B reaches A.

2 Particle P is initially at the point with position vector $4\mathbf{i} + 6\mathbf{j}$ and moves with velocity $3\mathbf{i} - 2\mathbf{j}$.

a) Find an expression for the position vector of P at any time t.

b) Find the time at which OP is parallel to the vector $5\mathbf{i} + \mathbf{j}$.

c) Find an expression for the distance of P from the origin at time t.

d) Find the minimum distance of P from the origin.

1a) Draw a diagram to work out the initial positions of A and B in vector form. Use the equation $\mathbf{r} = \mathbf{r}_0 + \mathbf{v}t$

1b) Use the equation $\mathbf{r} = \mathbf{r}_0 + \mathbf{v}t$ for each ship to find v. Remember speed is magnitude of velocity

1c) Find where each ship is at 9am using your equations from a). Find the distance between them and use speed = dist ÷ time

2a) Use the equation $\mathbf{r} = \mathbf{r}_0 + \mathbf{v}t$

2b) Look back at the example on parallel vectors

2c) Use the magnitude formula in the normal way

2d) When does what's in the square root take its smallest value?

10 mins

Time Yourself

Momentum

How much do you know?

1 a) A particle of mass 500 grams is moving with velocity $2\,\text{ms}^{-1}$. Find its momentum.

b) A particle of mass 2 kg is moving with velocity $6\mathbf{i} + 2\mathbf{j}$. It is acted on by a force $\mathbf{F}$ for 3 seconds, where $\mathbf{F} = -\mathbf{i} + 2\mathbf{j}$. Find its velocity after this time.

c) A particle of mass 1.5 kg is moving with speed $3\,\text{ms}^{-1}$ when it hits a wall. The wall exerts an impulse of 6 Ns on the particle. Find the speed with which the particle rebounds from the wall.

2 a) Particles P and Q have masses 1 kg and 3 kg respectively. Initially, P and Q are moving in the same directions with speeds $3\,\text{ms}^{-1}$ and $2\,\text{ms}^{-1}$ respectively. They collide and coalesce. Find the velocity with which the combined particle moves after the collision.

b) In an explosion, a motionless object of mass 20 kg splits into two pieces of masses 12 kg and 8 kg, which move in opposite directions. The 12 kg piece moves at a speed of $200\,\text{ms}^{-1}$. Find the speed of the 8 kg piece.

Answers

1a) $1\,\text{kgms}^{-1}$ b) $4.5\mathbf{i} + 5\mathbf{j}$ c) $1\,\text{ms}^{-1}$

2a) $2.25\,\text{ms}^{-1}$ b) $300\,\text{ms}^{-1}$

If you got them all right, skip to page 70

Spend no more than 45 mins on this topic

Learn the key facts

1 Impulse and Momentum

You find the momentum of a particle by multiplying its mass by its velocity. The important equation is:

momentum = mass × velocity

Momentum is measured in kgms^{-1}. Other units must be converted to kilograms, metres and seconds before you work out momentum.

You must always take care to take direction into account when you are working out momentum. It is best to decide to begin with that one direction is positive – so anything going in the opposite direction must have a negative speed.

You need to know about impulse, which causes a change in momentum. An object receives an impulse when:

- an external force acts on it for a period of time
- it hits another object which can move
- it hits a wall.

The important equation is:

impulse = force × time = new momentum – old momentum

Impulse is measured in the same units as momentum, or alternatively in Ns (Newton seconds) – again, make sure everything is in these units.

When two things hit each other, they exert equal and opposite impulses on each other

These two equations can also be applied if the forces or velocities are vectors (see the 'Vectors in Mechanics' chapter if you are unsure about using vectors).

Example 1

A particle of mass 6 kg is moving with velocity $4\mathbf{i} - 3\mathbf{j}$ metres per second. It is then acted on by a force $\mathbf{F}$ for four seconds, after which its velocity is $13\mathbf{j}$. Find:

a) the impulse which acts on the particle

b) the force $\mathbf{F}$.

Solution

a) impulse = new momentum − old momentum
$= 6(13\mathbf{j}) - 6(4\mathbf{i} - 3\mathbf{j})$
$= 78\mathbf{j} - 24\mathbf{i} + 18\mathbf{j}$
$= -24\mathbf{i} + 96\mathbf{j}$

b) force × time = impulse
$\mathbf{F} \times 4 = -24\mathbf{i} + 96\mathbf{j}$
$\mathbf{F} = -6\mathbf{i} + 24\mathbf{j}$

We get the momentum by multiplying mass by velocity

Example 2

A particle of mass 3 kg is travelling with speed 4 ms^{-1} when it hits a wall at right angles. When it rebounds from the wall, its speed is halved. Find the magnitude and direction of the impulse exerted by the wall on the particle.

Solution

We will take the direction the particle was originally moving in as positive. We can show the situation in a diagram:

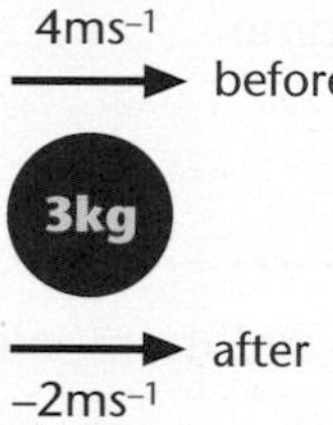

ALWAYS have the arrows going in the same direction – just put in a minus sign if you need it

Now use impulse = new momentum – old momentum:
$= 3 \times -2 - 3 \times 4 = -18\,\text{kgms}^{-1}$
So magnitude of impulse = 18 kgms^{-1}, in direction opposite to particle's initial direction of travel.

Always state the direction

2 Conservation of Momentum

When no external forces act, momentum is conserved. This gives use the equation:

total old momentum = total new momentum

You can't use this if particles hit a wall, since the wall provides an 'external force'.

Example 3

Two particles P and Q have masses 2 kg and 3 kg respectively and are travelling with velocities $4\,ms^{-1}$ and $3\,ms^{-1}$ respectively in the same direction on a smooth horizontal table. They collide and coalesce. Find their speed after the collision.

'Coalesce' means 'stick together'

Solution

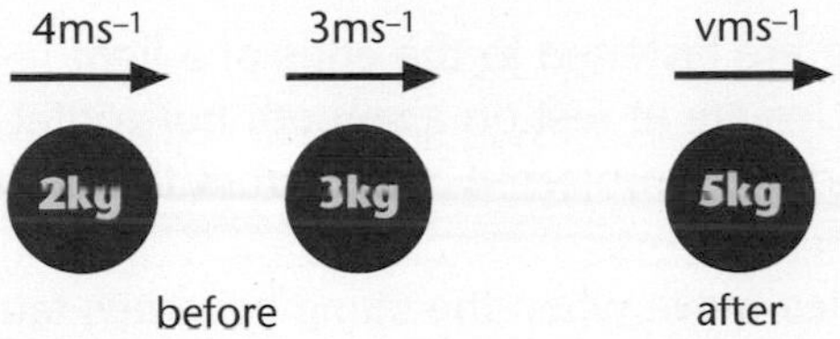

Total old momentum = total new momentum:

$2 \times 4 + 3 \times 3 = 5 \times v$

$17 = 5v \Rightarrow v = 3.4\ ms^{-1}$.

Example 4

A gun of mass 2 kg shoots a bullet of mass 50 g with a speed of $900\,ms^{-1}$. Find the speed of recoil of the gun.

Solution

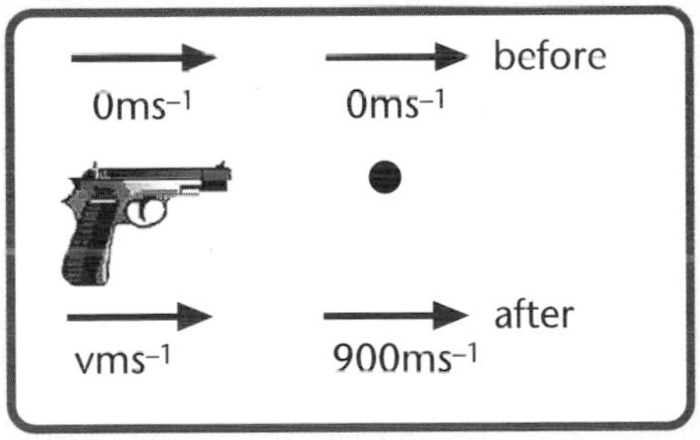

total old momentum = total new momentum

$0 = 0.05 \times 900 + 2 \times v$

$v = -22.5\,ms^{-1}$

So actual speed of recoil is $22.5\,ms^{-1}$.

30 mins
Time Yourself

Momentum

Have you improved?

1 Particle P, of mass 2m, is travelling with speed $3u$. It hits particle Q, which is of mass 3m and is initially stationary. Particle P is brought to rest in the collision. Find, in terms of u, the velocity with which particle Q moves initially.

1) Use conservation of momentum. Don't worry about it all being in terms of m and u – it will still work! Your answer should be in terms of u

2 Two particles of masses 1 kg and 2 kg are fastened to the ends of a light inextensible string. Initially, the particles are at rest on a smooth horizontal table next to each other. The 2 kg particle is projected away across the table with speed $6\,\text{ms}^{-1}$. Find:
a) the speed with which both particles move when the string becomes taut
b) the impulsive tension in the string as it becomes taut.

2a) Once the string is taut, the two particles must move with the same speed. Use conservation of momentum

2b) The impulsive tension in the string causes the 1kg particle to start moving. Use impulse= change in momentum for the 1 kg particle

3 Two particles P and Q have masses 2m and 3m respectively. They are initially moving in opposite directions with speeds $12u$ and $4u$ respectively. P and Q collide, and after the collision, P is moving with velocity $2u$ in the opposite direction. Find:
a) the velocity, in terms of u, of Q after the collision
b) the kinetic energy, in terms of m and u, lost in the collision.

3a) Conservation of momentum again. Be careful with signs and directions!

3b) The formula for kinetic energy is $\frac{1}{2}mv^2$. k.e. lost = total old k.e. – total new k.e.

4 A particle of mass 3 kg moving with speed $8\,\text{ms}^{-1}$ collides and coalesces with a particle of mass H kg, which is moving with speed $1\,\text{ms}^{-1}$ in the opposite direction. After the collision, the combined mass moves with speed $1.5\,\text{ms}^{-1}$ in the original direction of the 3 kg mass. Find H.

4) Conservation of momentum again. You will end up with an equation with H on both sides. Expand the brackets and simplify

5 Particles A, of mass 6m, and B, of mass 4m, are travelling with speeds $4\,\text{ms}^{-1}$ and $3\,\text{ms}^{-1}$ in the same direction. They collide, and after the collision, the velocity of B is $2\,\text{ms}^{-1}$ greater than the velocity of A. Find:
a) the speed of A after the collision
b) the impulse exerted by B on A during the collision.

5a) Call the new velocities v_A and v_B and use conservation of momentum. We must have $v_B = v_A + 2$. You will get simultaneous equations for v_A and v_B

5b) Use the equation for impulse for A

Descriptive Statistics

30 mins
Time Yourself

How much do you know?

1 a) Calculate the mean, mode, median, standard deviation and range for the following data: 3, 6, 8, 9, 8, 12, 17

b) Calculate the mean and standard deviation of the following data:

x	$0 \le x < 3$	$3 \le x < 5$	$5 \le x < 8$	$8 \le x < 10$
f	8	12	11	7

c) Explain what the effect would be on the mean and standard deviation if all the values in part a) were: i) increased by 6, ii) trebled.

2 Find the median and interquartile range of the following data:

x	1	2	4	5
f	6	6	5	1

3 Use linear interpolation to find: a) the median, b) the upper quartile, c) the 92nd percentile of the data in question 1b).

4 The following data were obtained for the times (in seconds) a class of 20 children took to run a race:

16.5	18.2	15.9	16.7	19.4	19.0	17.6	18.3	15.4	16.2
17.1	15.8	16.0	18.1	17.9	16.9	18.2	17.0	15.4	15.1

a) Draw a stem and leaf diagram to represent this data.

b) Draw a box plot to represent this data.

5

Test marks	Mean	Standard Deviation
Class A	65	8
Class B	68	4

a) Bhavanetta, who is in class B, said to Anna, who is in class A, 'That means we're all cleverer than you are.' Explain why the figures do not show this.

b) Which class do you think the person with the highest mark was in? Explain your reasoning.

Answers overleaf

Answers

If you got them all right, skip to page 80

1a) mean = 9, mode = 8, median = 8, SD = 4.1404, range = 14
b) Mean = 5.276, SD = 2.486
c) i) mean increased by 6, SD unaffected   ii) mean and SD both trebled
2 median = 2, IQR = 3
3a) 4.8333   b) 7.3182   c) 9.1314
4a)

15	1	4	4	8	9
16	0	2	5	7	9
17	0	1	6	9	
18	1	2	2	3	
19	0	4			

b)
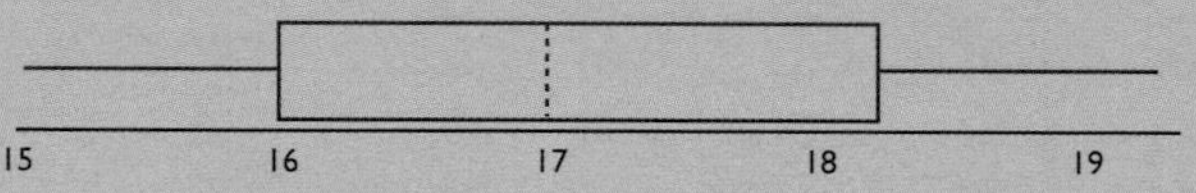

(LQ = 15.95; median =16.95; UQ= 18.15)

5a) Since the SD for class A is 8, there will be some people in class B who got below the mean of class A, since there is only 3 marks difference in the means. So *not all* of class B are cleverer. (Similarly, some people in A will have got above the mean of B.)
b) Since the marks in class A are much more variable, they are likely to have the highest mark (and the lowest).

Descriptive Statistics

Spend no more than 60 mins on this topic

Learn the key facts

1 Basic GCSE Calculations

From GCSE, you should know how to to calculate the mode, mean, variance, standard deviation and range. You should know how to use a cumulative frequency graph to find the median, quartiles and interquartile range. The formulae and facts you use are:

- Mode = most common value (or class)
- Mean = $\dfrac{\text{sum of the values}}{\text{number of values}}$ or $\dfrac{\sum fx}{\sum f}$ for frequency tables, where x is the mid-value of the group if it's a grouped frequency table.
- Variance = $\dfrac{\sum fx}{\sum f}$ – mean^2
- Standard deviation = √(variance)
- Range = highest value – lowest value
- Median: Using a cumulative frequency curve, look up $\frac{1}{2}n$ on y-axis (n = no. of values). The median is the corresponding point on the x-axis.
- Quartiles are calculated from a cumulative frequency graph like the median – the position of the LQ is given by $\frac{1}{4}n$, and of the UQ by $\frac{3}{4}n$.
- Interquartile range = upper quartile – lower quartile.

Σfx means 'add up all the fx values'

You also need to know from GCSE the effect on the mean and standard deviation if, for example, 4 is added to all the values, or all the values are doubled.

Learn how to find mean and SD on your calculator!

2 Median and Quartiles for Discrete Data

To find the median from a list of values, you put them in order and find the middle value. If there are an even number of values, there will be no real 'middle', so you average the two middle values.

To find the median and quartiles from an ungrouped frequency table, you adopt the following procedure:

1 Find the cumulative frequencies.

2 Find $\frac{1}{4}n$, $\frac{1}{2}n$ or $\frac{3}{4}n$ as appropriate, for the LQ, median or UQ.

3 You then find the position of the median or quartile as follows:

- If the number you have calculated in step 2 is <u>whole number</u>, then you average the value in this position and the value in the next position (e.g. if $\frac{1}{2}n = 5$, you'd average 5th and 6th values).
- If the number you have calculated in step 2 is <u>not a whole number</u>, then you go to the next value (e.g. if $\frac{1}{2}n = 5.5$, you'd go to the 6th value).

4 Once you have the position, you find the first value whose cumulative frequency is this number or higher (so if your median was in the 6th position, you'd choose the first value whose cumulative frequency was 6 or more).

Example 1

Find the median and interquartile range of the following data:

x	1	2	4	6	7
f	2	4	7	8	5

Solution

First find cumulative frequencies:

x	f	c.f.
1	2	2
2	4	6
4	7	13
6	8	21
7	5	26

Median: $\frac{1}{2}n = \frac{1}{2} \times 26 = 13$

So we want average of 13th and 14th values.
To find 13th value: c.f. is first 13 or higher when $x = 4$. So 4 is 13th value.
To find 14th value: c.f. is first 14 or higher when $x = 6$. So 6 is 14th value.
So median is average of 4 and 6 = 5

LQ: $\frac{1}{4}n = \frac{1}{4} \times 26 = 6.5$

So we want 7th value.
c.f. is first 7 or higher when $x = 4$. So 4 is 7th value.
So LQ = 4

UQ: $\frac{3}{4}n = \frac{3}{4} \times 26 = 19.5$

So we want 20th value.
c.f. is first 20 or higher when $x = 6$. So 6 is 20th value.
So UQ = 6
So the interquartile range = 6 – 4 = 2.

3 Median and Quartiles for Grouped Data by Linear Interpolation

If you have grouped data, you are more likely to be asked to find median and quartiles by linear interpolation than by drawing a cumulative frequency graph. Linear interpolation gives you the answer you'd get if you drew a perfectly accurate cumulative frequency polygon!

To carry it out, you go through these steps:

1. Find the true class boundaries. For example, if the table referred to weights to the nearest kilogram, then the class 4 kg – 6 kg would actually include weights 3.5 kg–6.5 kg.
2. Calculate the cumulative frequencies and $\frac{1}{4}n$, $\frac{1}{2}n$, $\frac{3}{4}n$ as appropriate.
3. Decide which class the median or quartile is in by looking at cumulative frequencies.
4. Use the formula:

$$\text{median} = \text{lower class boundary} + \frac{\frac{1}{2}n - \text{c.f. of class below}}{\text{class frequency}} \times \text{class width}$$

(Class width = upper class boundary – lower class boundary)

The formula for the LQ and UQ has $\frac{1}{2}n$ replaced by $\frac{1}{4}n$, or $\frac{3}{4}n$.

You can also use this method to find percentiles. For the kth percentile, you replace $\frac{1}{2}n$ in the formula by $\frac{k}{100}n$.

Example 2

Find the median, the lower quartile and the 8th percentile of the following data:

Length (nearest cm)	1–3	4–8	9–11	12–15	16–26	27–50
Frequency	5	6	8	9	6	3

Solution

First, find true class boundaries and cumulative frequencies:

x	f	c.f.
0.5–3.5	5	5
3.5–8.5	6	11
8.5–11.5	8	19
11.5–15.5	9	28
15.5–26.5	6	34
26.5–50.5	3	37

If you've done the boundaries correctly, there'll be no overlaps or gaps in your classes

Median: $\frac{1}{2}n = \frac{1}{2} \times 37 = 18.5$. This is in the class 8.5–11.5

So median $= 8.5 + \frac{18.5 - 11}{8} \times (11.5 - 8.5) = 11.3125$

Lower quartile: $\frac{1}{4}n = \frac{1}{4} \times 37 = 9.25$. This is in the class 3.5–8.5

So lower quartile $= 3.5 + \frac{9.25 - 5}{6} \times (8.5 - 3.5) = 7.0417$ (4DP)

8th percentile: $0.08n = 2.96$. This is in the class 0.5–3.5

So 8th percentile $= 0.5 + \frac{2.96 - 0}{5} \times (3.5 - 0.5) = 2.276$

As there is no class below, the c.f. for class below is zero

4 Diagrams

From GCSE Higher, you need to be able to draw cumulative frequency curves and polygons, histograms and frequency polygons. You now also need to be able to produce stem and leaf diagrams and box plots (or box and whisker plots). A box plot must always be done on graph paper with a scale. The diagram below shows you what to put on it:

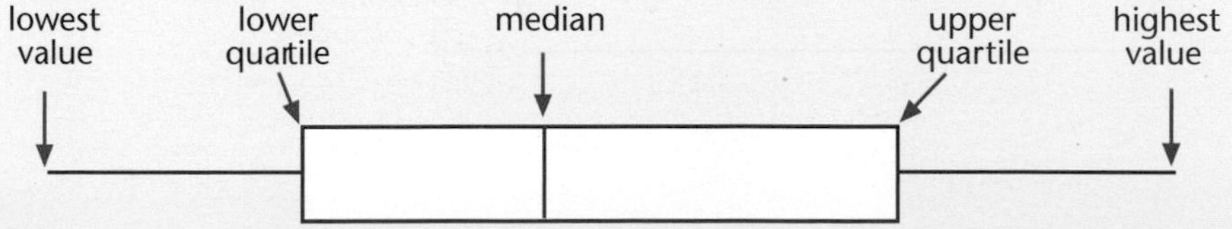

In a stem and leaf diagram, the 'leaves' are the final digits of the data values and the 'stems' are the rest of the value. The 'leaves' must be arranged in order, and aligned under one another. It is usually helpful to do a rough version first, where the leaves are not aligned.

Below is shown a stem and leaf diagram for the following data:

24 26 54 31 47 36 30 38 21 29 48 52 44 40 41

Stem	Leaves				
2	1	4	6	9	
3	0	1	6	8	
4	0	1	4	7	8
5	2	4			

Advantages and Disadvantages of Types of Diagram and When They Can be Used

This commonly comes up as the end of a question!
The table summarises what to learn.

Diagram	When you can use it	Advantages	Disadvantages
Histogram	Continuous data	• Allows you to see overall distribution of data – where the peak is, etc.	• Actual data values are lost
Stem & Leaf	Discrete data	• All the data values are retained • Makes it easy to find median & quartiles • You can still get an overall picture of distribution • You can use it to compare two distributions	• Only practicable for a small amount of data
Box plot	Discrete or continuous	• Allows you to see the position of the 'middle 50%' of data easily • Allows you to compare distributions	• Actual data values are lost • It does not give a detailed picture of the shape of the distribution

5 Interpreting and Commenting on Diagrams and Calculations

You need to know what all the calculations tell you!

- The mean, median and mode are all 'averages' – the mean is the 'traditional' average; the median has half the values above it and half below; and the mode is the commonest – or 'typical' value.
- The mode is not used much in statistics, because it may not exist – two or more values may be equally common.
- The interquartile range, standard deviation and range are all measures of how spread out the data are – if they are large, the data are more spread out. This is useful for comparing two sets of data.
- If you have a few unusual, high values, they pull the mean and SD up so that they become 'uncharacteristic' of the data – in a case like this the median and IQR are more useful.

If you are asked to comment, bear in mind the following:

- Comparisons must make reference to the figures given in the question and/or the diagrams you have drawn if you want any marks!
- You must make (at least) as many points in your comparison as there are marks for the question.
- You must use the right words – e.g. make it clear which 'average' you are talking about.
- You can comment on: comparison between one or more averages, the range of the data and the spread of the data.
- Your comments should take into account the context of the question – refer to what you mean in real life.

Example 3

A survey on lateness of trains gathered the data shown in the table from trains at New Hill station and Snow Street station in August 1998. The figures refer to lateness of trains in minutes.

	Lower Quartile	Median	Upper Quartile	Mean	Mode
New Hill	6	12	27	16	10
Snow Street	8	13	22	14	11

a) James said: 'Trains from Snow Street are generally later than those from New Hill.'
   i) Give some reasons to back up James' point of view.
   ii) Give some reasons why someone may disagree with James.
b) Compare the performance of the trains at the two stations.
c) One of the stations had engineering works taking place that affected some trains, but not all, in this period. Which station do you think this was? Explain your answer.

Solution

a) i) Snow Street has a higher median, lower quartile and mode.
   ii) New Hill has a higher upper quartile and mean.
b) The fact that New Hill has a lower median and mode suggests that its relatively high mean is produced by a few exceptionally late trains, and that in fact, most trains are earlier at New Hill. This is reinforced by the fact that lateness at New Hill is more spread out (IQR for New Hill is 21, whereas at Snow Street it is only 14), and that the upper quartile at New Hill is large, indicating that the worst 25% of trains are significantly later than at Snow Street. So, to summarise, the trains are more consistent at Snow Street (since IQR is smaller and mean, median and mode are relatively close), whereas at New Hill, while many trains are actually earlier than at Snow Street, a few are substantially later.
c) New Hill, since this ties in with a few trains being much later than at Snow Street.

Descriptive Statistics

Have you improved?

1 The data opposite represents the weights of some students (to the nearest kilogram).

a) Find an estimate of the mean and standard deviation of the weights, and explain why it is only an estimate.

b) Find the median and interquartile range of the weights.

c) Compare the mean and the median.

Weight (kg)	Number of people
40–49	2
50–54	10
55–59	30
60–64	40
65–69	15
70–89	3

1a) Since it is measured to the nearest kg, you need to work out the real lower and upper boundaries for each class. Use the class midpoints as your x values. Do you know the real data values?

1b) You need to use linear interpolation

1c) Are the values close? What does this tell you? Has the mean been distorted?

2 a) Calculate the mean and standard deviation of: 1, 2, 3, 4, 5.

b) Hence obtain the mean and standard deviation of:

i) 10, 11, 12, 13, 14

ii) 2, 4, 6, 8, 10

iii) 32, 34, 36, 38, 40

iv) $x + y, 2x + y, 3x + y, 4x + y, 5x + y$

2a) Look up the formulae in part 1

2b) Refer to GCSE work! The mean is affected in the 'obvious' way. The SD is not affected by addition/subtraction, only by multiplication/ division

3 The table below shows the ages (in completed years) of a random sample of 60 students at Notown Metropolitan University.

Age	16–17	18	19	20	21–25	26–40
Frequency	2	19	17	16	3	3

a) Explain why a histogram is suitable to represent this data.

b) Give one advantage of using a histogram to represent this data.

The 'bar' on the histogram for the age 21–25 is of width 2.5 cm and height 1.2 cm.

c) Find the heights and widths of the bars representing:

i) age 16–17 ii) age 20.

3a) Is age really continuous or discrete?

3b) Look back at the table in section 5

3c) Think carefully about class boundaries – you are 25 right up until your 26th birthday. Use the width to work out how many cm for one year. Work out the frequency density for the 21–25 class, and then you can find the vertical scale

Probability

30 mins *Time Yourself*

How much do you know?

1 A student shops at Kwikways 30% of the time, and Safesave the rest of the time. When he goes to each shop, he buys either a can of beans or a pizza. At Kwikways, he buys beans 40% of the time, but at Safesave he buys beans 65% of the time.

a) Find the probability that on a randomly selected shopping trip he buys a pizza.

b) Hence find the probability that he buys beans.

2 When I go to work, I catch two buses. My first bus is late 25% of the time. My second bus is late 30% of the time. I am only late for work if both buses are late; this happens 10% of the time.

a) Find the probability that at least one bus is late.

b) Find the probability that neither bus is late.

3 Tegan always either goes to the cinema or goes to the disco on Saturday evenings, but never both. She goes to the cinema 60% of the time. If she goes to the cinema, she will have a lie-in on Sunday morning with a probability of 0.4; if she goes to the disco, she has a lie-in with probability 0.7.

a) Find the probability Tegan has a lie-in on a randomly selected Sunday.

b) Given that Tegan did not have a lie-in on Sunday, find the probability she went to the disco on Saturday evening.

4 a) I throw a fair dice. Event A occurs if I get a prime number. Event B occurs if I get a square number. Event C occurs if I get an even number.

i) Find the probabilities of events A, B and C.

ii) Show that events A and B are mutually exclusive, and that events B and C are independent.

iii) Are events A and C independent? Justify your answer.

b) Given that $P(A) = 0.6$, $P(B) = 0.3$, and that events A and B are independent, find:

i) $P(A \cap B)$ ii) $P(A|B)$ iii) $P(A \cup B)$.

Answers

1a) 0.425 b) 0.575 **2**a) 0.45 b) 0.55 **3**a) 0.52 b) 0.25

4a) i) $P(A) = \frac{1}{2}$ $P(B) = \frac{1}{3}$ $P(C) = \frac{1}{2}$ ii) A and B cannot happen at once because A only happens if I get 2, 3, 5, and B only happens if I get 1, 4.

$P(B \cap C) = P(4) = \frac{1}{6}$. $P(B) \times P(C) = \frac{1}{3} \times \frac{1}{2} = \frac{1}{6}$. So B and C are independent.

iii) $P(A \cap C) = P(2) = \frac{1}{6}$. $P(A) \times P(C) = \frac{1}{2} \times \frac{1}{2} = \frac{1}{4}$ So A and C are not independent

b) i) 0.18 ii) 0.6 iii) 0.72

If you got them all right, skip to page 86

Spend no more than 45 mins on this topic

Probability

Learn the key facts

1 Tree Diagrams

From GCSE, you need to know how to use tree diagrams. The tree diagrams you will meet at AS Level will usually have the second lot of branches depending on the first, as in Example 1.

When using a tree diagram, multiply along branches then if you need more than one branch, add them

Example 1

I have a bag containing 8 red sweets, 3 green sweets and 2 orange sweets. I take 2 sweets out of the bag. What is the probability:
a) they are the same colour b) they are of different colours?

Solution

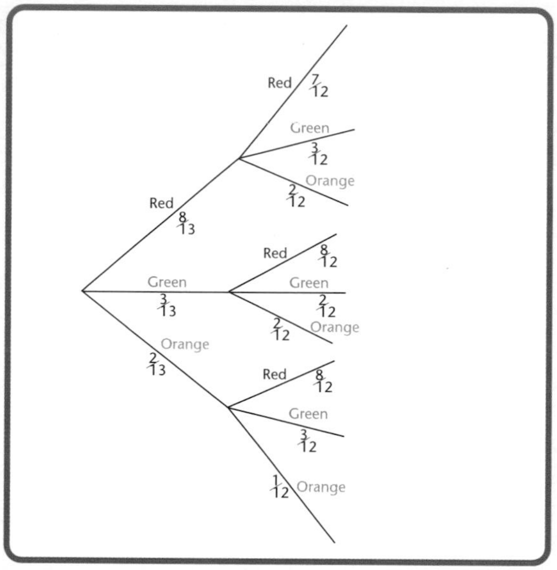

a) To get the same, we need RR, GG or OO.

$P(RR)= \frac{8}{13} \times \frac{7}{12} = \frac{14}{39}$; $P(GG) = \frac{3}{13} \times \frac{2}{12} = \frac{1}{26}$; $P(OO)= \frac{2}{13} \times \frac{1}{12} = \frac{1}{78}$

So P(both the same) $= \frac{14}{39} + \frac{1}{26} + \frac{1}{78} = \frac{16}{39}$

b) Since this is the opposite of both sweets being the same,
P(different) $= 1 - $ P(same) $= 1 - \frac{16}{39} = \frac{23}{39}$

This is the 'NOT' rule. Remember it!

DAY 6

2 Probability Notation and Formulae

You need to be able to recognise and use probability notation:

- $P(A \cap B)$ means the probability A and B both happen.
- $P(A \cup B)$ means the probability A or B or both happen.
- $P(A')$ means the probability A doesn't happen ($= 1 - P(A)$).
- $P(B|A)$ means the probability B happens given A has happened (or the probability of B conditional on A).

'A or B' usually means A or B or both in probability

You also need to be able to use the following formulae, which should be in your formula book:

$$P(A \cup B) = P(A) + P(B) - P(A \cap B); \quad P(B|A) = \frac{P(B \cap A)}{P(A)}$$

Example 2 shows you how to use the first formula. See the next section for help on the second one.

Example 2

The probability I have a dessert with my lunch is 0.2. The probability I have a dessert with my dinner is 0.4. On 10% of days I have desserts with both meals. On what percentage of days do I have: a) at least one dessert; b) no dessert?

'at least one' means 'one or the other or both'

Solution

Let dessert with lunch = L. Let dessert with dinner = D. So $P(L) = 0.2$; $P(D) = 0.4$

We known $P(\text{both}) = 0.1$; 'both' means $L \cap D$, so $P(L \cap D) = 0.1$

a) We want P(at least one dessert) = P(lunch or dinner or both) = $P(L \cup D)$
Using the formula: $P(L \cup D) = P(L) + P(D) - P(L \cap D)$
$= 0.2 + 0.4 - 0.1 = 0.5$ – which is 50% of days.

b) 'No dessert' is the opposite of 'at least one dessert'.
So P(no dessert) $= 1 - 0.5 = 0.5$ – giving 50% again.

It's useful to remember: P(at least one) = 1 − P(none)

3 Using Conditional Probability

In section 2 you were given a formula for conditional probability. Conditional probability is about the chance something happens if you know something else has already happened – for example, if you know your friend has forgotten his homework, that will affect the probability he gets in trouble with the teacher!

You know you have to use conditional probability if you see any of the following in a question:

- 'Find the conditional probability that ...'
- 'Given that ... find the probability that ...' or 'Find the probability that ... given that ...'
- You know something about what's happened already.

When you write P(B|A), the thing that you know, or that you are given, is A and the thing you want to find out is B. Example 3 shows how to use it.

Example 3

Rena travels to school by bus 40% of the time, and the rest by bike. When she travels by bus, the probability of her being late is 0.3; when she travels by bike, this probability is 0.2.

a) Find the probability she is late.

b) Given she is late, what is the probability she used the bus?

Solution

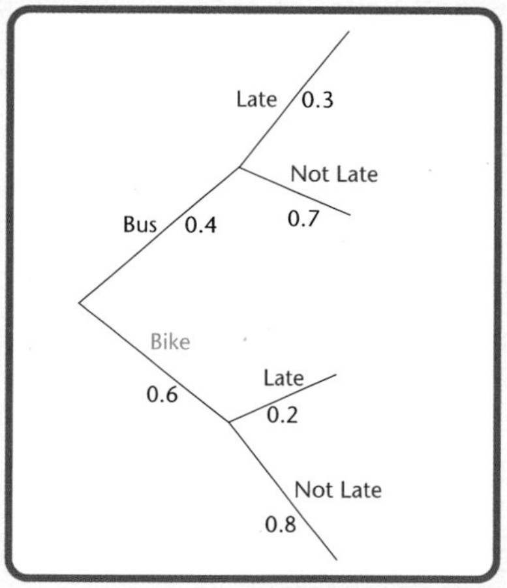

a) $P(\text{Bus and late}) = 0.4 \times 0.3 = 0.12;$
$P(\text{bike and late}) = 0.6 \times 0.2 = 0.12$
$\Rightarrow P(\text{late}) = 0.12 + 0.12 = 0.24$

b) This is $P(\text{bus}|\text{late}) = \dfrac{P(\text{bus and late})}{P(\text{late})} = \dfrac{0.12}{0.24} = 0.5$

DAY 6

4 Independent and Mutually Exclusive Events

- Independent events do not affect each other.
  For independent events, $P(A \cap B) = P(A) \times P(B)$
- Mutually exclusive events can't both happen.
  For mutually exclusive events, $P(A \cap B) = 0$

If a question tells you that events are independent/mutually exclusive, then use the appropriate formula from above.
If a question asks you to show that events are independent/mutually exclusive, then you check to see if the appropriate formula works. If it does not, then the events are not independent/mutually exclusive.

Example 4

$P(A \cup B) = 0.8$; $P(A) = 0.5$
Find $P(B)$ if: a) A and B are mutually exclusive; b) A and B are independent.

Solution

a) We know $P(A \cap B) = 0$. Since we are given $P(A \cup B)$, it makes sense to use
$P(A \cup B) = P(A) + P(B) - P(A \cap B)$
So $0.8 = 0.5 + P(B) - 0$.
So $P(B) = 0.3$

b) We know $P(A \cap B) = P(A) \times P(B) = 0.5P(B)$
Again, use $P(A \cup B) = P(A) + P(B) - P(A \cap B)$:

$$0.8 = 0.5 + P(B) - 0.5P(B) \Rightarrow 0.3 = 0.5P(B) \Rightarrow P(B) = \frac{0.3}{0.5} = 0.6$$

Example 5

$P(A \cup B) = 0.7$; $P(A) = 0.5$; $P(B) = 0.3$. Investigate whether A and B are independent.

Solution

To see whether A and B are independent, we must check whether
$P(A \cap B) = P(A) \times P(B)$ or not.

We therefore need to find $P(A \cap B)$.
Since we are given $P(A \cup B)$, use $P(A \cup B) = P(A)+P(B) - P(A \cap B)$:
$0.7 = 0.5 + 0.3 - P(A \cap B) \Rightarrow P(A \cap B) = 0.1$

$P(A) \times P(B) = 0.5 \times 0.3 = 0.15$; $P(A \cap B) = 0.1$
So $P(A) \times P(B) \neq P(A \cap B) \Rightarrow$ A and B not independent.

30 mins
Time Yourself

Probability

Have you improved?

1 0.5% of the population suffer from a rare disease. There is a test for this disease; it gives a positive result with 99% of people who have the disease, and with 2% of people who do not have the disease.

a) Find the probability that a randomly selected individual will test positive for the disease.

b) Shirley has just tested positive for the disease. Find the probability she actually has the disease.

c) Comment on your answer to part b).

2 a) Explain why $P(A \cap B) + P(A \cap B') = P(A)$.

In a certain school, all girls play either hockey or netball or both.
Given that P(netball|hockey) = $\frac{1}{4}$, P(netball) = $\frac{11}{20}$ and P(hockey) = h:

b) Explain why P(both sports) + P(just netball) + P(just hockey) = 1.

c) Show P(just hockey) = $\frac{9}{20}$.

d) Show that $h = \frac{9}{20} + \frac{1}{4}h$, and hence find the value of h.

e) Find P(not hockey|netball).

1a) Draw a tree diagram. Don't forget 0.5% is not a probability of 0.5

1b) You already know something. What does that tell you about the sort of question it is?

1c) What proportion of people with positive tests are actually ill?

2a) Think what $A \cap B$ and $A \cap B'$ mean. P(X) + P(Y) = P(X or Y) if X and Y can't happen at once

2b) Think what the different alternatives are.

2c) Write out part b) in symbols, remembering 'just hockey' means 'hockey and not netball'. Use the result given in part a), with A = netball

2d) Use the result given in part a), with A = hockey. Use the conditional formula to get P(hockey and netball)

2e) Use the result given in part a), with A = netball. Work out P(hockey and netball), now you know h.

Correlation & Regression

20 mins
Time Yourself

How much do you know?

1 Mark wanted to find out whether there was a correlation between the mileage and cost of a 'Vienna' car. He collected the following data from nine second-hand car dealers in Birmingham.

Mileage, x (in 000's)	11.0	26.3	30.1	40.0	46.1	54.8	55.3	71.0	88.0
Price, y £ (in 000's)	13.4	8.2	10.9	8.9	6.1	5.1	7.2	4.0	2.3

(You may assume that $\sum x^2 = 24290.04$, $\sum y^2 = 581.17$ and $\sum xy = 2192.4$.)

a) Calculate the value of the product moment correlation coefficient.
b) Comment on the value of r that you have found in (a).

2 The following table shows the lengths (in mm) and the widths (in mm) of ten randomly collected ivy leaves.

Length, x (mm)	40	45	54	58	77	82	87	98	107	117
Width, w (mm)	26	33	29	41	57	48	58	58	72	86

(You may use $\sum x^2 = 64869$, $\sum w^2 = 29148$ and $\sum xw = 43290$.)

a) Find the equation of the regression line w on x in the form $w = a + bx$.
b) Interpret the value of the coefficient b.
c) Using the equation of the regression line calculated, find estimates for the width of an ivy leaf when its length is: i) 60 mm, ii) 25 mm.
Comment on the reliability of each estimate.

Answers

1a) −0.937   b) Value of r is close to −1 ⇒ from this sample we can say that there is a strong negative linear correlation between the mileage and the price of a second-hand 'Vienna' car.
2a) $w = -2.57 + 0.698x$ b) For every extra increase of 1 mm in the length of a leaf, its width increases by a further 0.698 mm.   c)(i) 39.3 mm, reliable estimate because its been estimated from within the range of lengths given. ii) 14.9 mm, unreliable because we have extrapolated (i.e. found an estimate outside the range of the data given).

If you got them all right, skip to page 93

Spend no more than **35 mins** on this topic

Correlation & Regression

Learn the key facts

1 Product Moment Correlation Coefficient (PMCC)

The product moment correlation coefficient (PMCC) is a numerical measure of the degree of linear *relationship* between two variables. The PMCC is measured on a scale from −1 to +1. The higher the value of r (ignoring the negative sign), the closer the scatter points lie to a straight line.

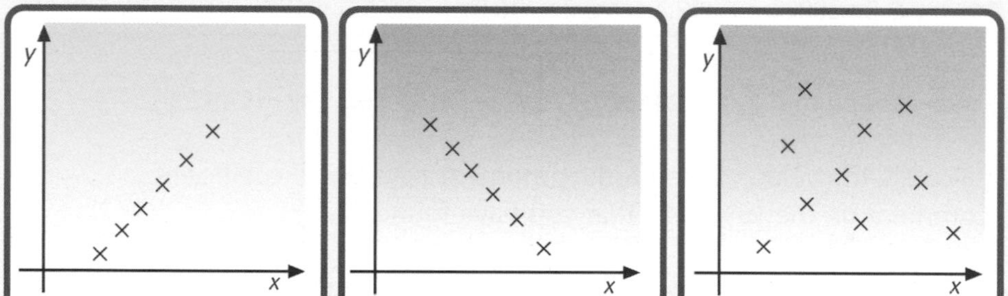

$r = 1$ perfect positive linear correlation | $r = -1$ perfect negative linear correlation | $r \simeq 0$ no linear correlation

In AS Statistics, we are only concerned with linear correlation. If $r \approx 0$, it may be possible that the variables could be related in a non-linear way (e.g. quadratic).

To calculate r, we use a formula: $r = \dfrac{S_{xy}}{\sqrt{S_{xx}.S_{yy}}}$

where: $S_{xy} = \sum xy - \dfrac{\sum x \sum y}{n}$ $\quad S_{xx} = \sum x^2 - \dfrac{\left(\sum x\right)^2}{n}$ $\quad S_{yy} = \sum y^2 - \dfrac{\left(\sum y\right)^2}{n}$

Example 1

A maths teacher wants to find out whether there is any relationship between GCSE examination marks and the coursework marks for his students. He obtained the following results for eight randomly selected students:

Student:	**A**	**B**	**C**	**D**	**E**	**F**	**G**	**H**
Exam mark, x	57	69	89	35	60	63	50	77
Coursework mark, y	25	29	35	15	23	30	28	32

a) Calculate S_{xy}, S_{xx} and S_{yy}.
b) Hence find the product moment correlation coefficient for the data.

The maths teacher said: 'Students who perform well in their maths exam also do well in their coursework'.
c) Comment on the maths teacher's statement.

Solution

a) Draw a table with columns: x, y, x^2, y^2 and xy, which are then summed.

	x	y	x^2	y^2	xy
	57	25	3249	625	1425
	69	29	4761	841	2001
	89	35	7921	1225	3115
	35	15	1225	225	525
	60	23	3600	529	1380
	63	30	3969	900	1890
	50	28	2500	784	1400
	77	32	5929	1024	2464
$\Sigma =$	500	217	33154	6153	14200

Hence, using the table:

$$S_{xy} = 14200 - \frac{500 \times 217}{8} = 637.5$$

$$S_{xx} = 33154 - \frac{(500)^2}{8} = 1904$$

$$S_{yy} = 6153 - \frac{(217)^2}{8} = 266.875$$

b) $$r = \frac{S_{xy}}{\sqrt{S_{xx}.S_{yy}}} = \frac{637.5}{\sqrt{1904 \times 266.875}} = 0.894...(3\,sf)$$

c) Since the value of r calculated is fairly close to $+1$, we can support the teacher's statement.

1 2 3 4 5 6 DAY 7

2 Linear Regression

A least squares (LS) regression line is the equation of the 'line of best fit'. The LS regression line has equation $y = \alpha + \beta x$, where α is the y-intercept and β is the gradient.

- x: independent (explanatory) variable
- y: dependent (response) variable
- only use x-values to predict y-values
- an LS regression line is derived so that the total vertical distances (errors) from the actual y-values (scatter points) to the predicted y-values (on the regression line) are minimised
- all LS regression lines go through $(\bar{x}, \bar{y})$.

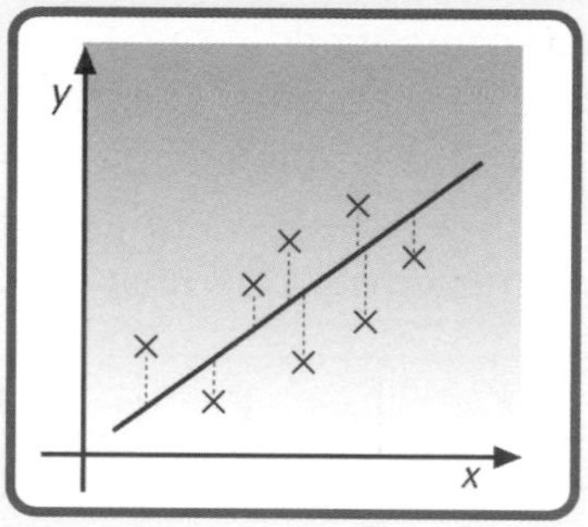

To calculate the LS regression line y on x, we use the formula:

$$y = \alpha + \beta x, \text{ where } \beta = \frac{S_{xy}}{S_{xx}} \text{ and } \alpha = \bar{y} - \beta \bar{x}$$

Some questions may ask you to give an interpretation of α and/or β. You will not get any marks for saying α is the y-intercept and β is the gradient. You must say:

α: When the [*insert what x means*] is 0, the value of [*insert what x means*] is α.

$\beta > 0$: For every extra one unit of [*insert what x means*], the [*insert what x means*] increases by β.

$\beta < 0$: For every extra one unit of [*insert what x means*], the [*insert what x means*] decreases by β.

Example 2

Jacob was conducting an experiment to find out the whether there is a linear relationship between force applied to a spring (in Newtons) and its extension (in centimetres). Jacob applied forces of 0 N, 10 N, 20 N, 30 N, etc. to the spring and measured how far the spring had extended from its natural length. His results are recorded in the table below:

Force applied (N); *x*	0	10	20	30	40	50	60	70
Extension (cm); *y*	0	1.3	2.4	3.7	4.9	6.4	7.8	9.2

(You may assume that: $\Sigma x^2 = 14000$, $\Sigma xy = 1800$, $\Sigma x = 280$, $\Sigma y = 35.7$.)

a) Plot a scatter diagram of y against x.
b) Find S_{xx} and S_{xy}.
c) Calculate the equation of the regression line y on x, in the form $y = a + bx$.
d) Give an interpretation of the constant b.

e) Draw the regression line on your graph.
f) Using the equation of the regression line calculated, find an estimate for the extension of the spring when the force applied is:
i) 25 Newtons ii) 150 Newtons.
Hence comment on the reliability of your estimates.

Bryan says that he prefers the extension to be measured in millimetres.
g) Derive the equation of the regression line z on x, where z is the extension in millimetres.

Solution

a)

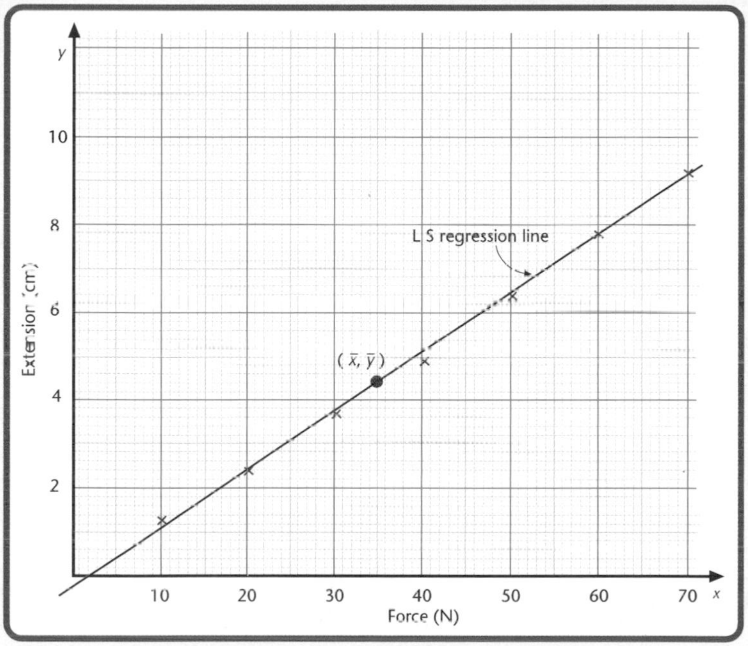

When drawing your scatter diagram, make sure you label both the x- and y-axes.

b) $S_{xx} = 14000 - \dfrac{(280)^2}{8} = 4200$ $\qquad S_{xy} = 1800 - \dfrac{280 \times 35.7}{8} = 550.5$

c) Find b: $b = \dfrac{S_{xy}}{S_{xx}} = \dfrac{550.5}{4200} = 0.13107...$

- Means: $\bar{x} = \frac{\sum x}{n} = \frac{280}{8} = 35$ and $\bar{y} = \frac{\sum y}{n} = \frac{35.7}{8} = 4.4625$

- Find a: $a = \bar{y} - b\bar{x} = 4.4625 - (0.13107\ldots)(35) = -0.125$

- $y = a + bx \Rightarrow y = -0.125 + 0.131x$ **(1)**

d) b: For every extra 1 Newton force added to the spring, the extension increases by a further 0.131 cm.

e) Plot the points $(0, -0.125)$ and $(\bar{x}, \bar{y}) = (35, 4.5)$ on the scatter diagram, and join these two points up with a straight line.

f) i) $y = -0.125 + 0.131(25) = 3.15$ cm; reliable estimate because the scatter points lie close to a straight line and $x = 25$ N is within the range of the data.
ii) $y = -0.125 + 0.131(150) = 19.5$ cm; not reliable because we do not know that a linear relationship exists beyond $x = 70$ N (i.e. we are extrapolating).

g) To convert centimetres to millimetres, we multiply by 10; i.e.
$10y = z \Rightarrow y = \frac{z}{10}$.

Substituting into **(1)** gives: $\frac{z}{10} = -0.125 + 0.131x \Rightarrow z = -1.25 + 1.31x$.

In this example the forces applied to the spring (x-values) ranged from 0 N to 70 N. When we drew the scatter diagram, we observed from the scatter points that a linear relationship was valid. This linear relationship is only valid for values of x within the range from 0N to 70N. We have no evidence about what happens to the spring after we have applied a force of 70N to it.

In f) i), when $x = 25$N, the estimate was reliable because 25N is within the range of the experimental results. The process we have used to get this estimate is interpolation.

However, in f) ii), when $x = 150$N, the estimate is unreliable because $x = 150$N is far outside the range of data. The process used to get this estimate is extrapolation. Estimates found by use of extrapolation should always be treated with caution.

30 mins
Time Yourself

Have you improved?

1 Sangita wanted to find out if there is any relationship between a person's weight and their systolic blood pressure. She took a random sample of 15 people, recording their weight, x (in kg), and blood pressure, y (in mmHg). She calculated the average weight to be 97.3 kg and the average blood pressure as 138.2 mmHg. Using her calculator she found: $\sum x^2 = 146587$, $\sum y^2 = 289111$ and $\sum xy = 202009$.

a) Calculate the value of the product moment correlation coefficient, leaving your answer to 3 decimal places.

b) Comment on the value of r that you have found.

1a) Find Σx and Σy. Find the Ss then r

1b) Refer to the question

2 The yield (in grams) from a chemical reaction and the temperature (in °C) are known to have an approximate linear relationship for a certain range of temperatures. The table below shows yields obtained at 8 different temperatures.

Temp, x (°C)	80	90	100	110	120	130	140	150
Yield, y (g)	52	63	66	54	76	84	87	95

a) Plot a scatter diagram to represent the data.

b) It transpired that the measuring scales were faulty for one of the readings. State, giving a reason, the temperature at which this occurred.

c) Ignoring the defective measurement, calculate the equation of the regression line y on x.

2a) On graph paper

2b) Use your scatter diagram to decide

2c) Find β, then α. Then use $y = \alpha + \beta x$

By adding a special chemical to the reaction, each of the yields increases, with the temperature remaining unchanged.

d) Find the equation of the new regression line if all yields had increased by 15% in mass.

2d) $y_{new} = 1.15\, y_{old}$

3 Give a practical interpretation of the coefficients α and β in $y = \alpha + \beta x$, for the following cases:

a) $y = 3.75 + 0.28x$, where y represents the weight of a baby in kg and x represents its age in months.

b) $y = 230 - 7.5x$, where y represents the units of gas used and x represents the temperature outside in °C.

3a) α: when $x = 0$, $y = ?$

3b) Refer to the questions in your answer!!

1 2 3 4 5 6 7

30 mins
Time Yourself

Discrete Random Variables

How much do you know?

1 A discrete random variable, Y, has the following probability distribution:

y:	**−2**	**−1**	**0**	**1**	**2**
p(y):	k	$\frac{1}{24}$	$\frac{1}{8}$	k	$\frac{1}{3}$

Find:
a) the value of k
b) $P(Y > -1)$
c) $P(Y \leq 0)$
d) $P(Y > -0.3)$
e) $P(-1 < Y \leq 1)$
f) $P(-1 \leq Y \leq 1)$
g) $F(-1)$
h) $F(0.2)$

2 The random variable X has the probability distribution:

x:	1	2	3	4	5
p(x):	0.05	0.15	0.4	0.3	0.1

Find:
a) $E(X)$
b) $E(X^2)$
c) $Var(X)$
d) the standard deviation of X

3 You are given that: $Var(X) = 5$ and $E(X) = 2$. Find:
a) $E(X^2)$
b) $E(2X - 1)$
c) $E(X + 3)$
d) $E(2 - 3X)$
e) $Var(4X)$
f) $Var\left(\frac{X}{3}\right)$
g) $Var(2X - 3)$
h) $Var\left(\frac{X+5}{2}\right)$

4 A bag contains 4 green, 4 blue and 2 black balls. Two balls are selected at random from the bag without replacement. The random variable Y is defined as the number of green balls selected.
a) Find the probability distribution of Y.
b) Hence find the mean number of green balls selected.

Answers

1a) $\frac{1}{4}$ b) $\frac{17}{24}$ c) $\frac{5}{12}$ d) $\frac{17}{24}$ e) $\frac{3}{8}$ f) $\frac{5}{12}$ g) $\frac{7}{24}$ h) $\frac{5}{12}$
2a) 3.25 b) 11.55 c) 0.9875 d) 0.994...
3a) 9 b) 3 c) 5 d) −4 e) 80 f) 5/9 g) 20 h) 1.25
4a) y: 0 1 2 b) 0.8
$p(y)$: $\frac{1}{3}$ $\frac{8}{15}$ $\frac{2}{15}$

If you got them all right, skip to page 98

DAY 7

Discrete Random Variables

Spend no more than 60 mins on this topic

Learn the key facts

1 Random Variables

- A random variable, X, is a variable whose values, x, are determined with specified probabilities.
- A *discrete random variable* takes set values (usually integers), while a *continuous random variable* contains values in a range.
- A *probability distribution* is the set of values that x can take, with their corresponding probabilities.

You need to be learn how to use the following formulae:
$\Sigma p(x) = 1$, i.e. *all probabilities add up to make 1.* (NB: $p(x) = P(X = x)$)
$F(x) = P(X \leq x)$: F is the cumulative distribution function.

Example 1

A discrete random variable, X, has the following probability distribution:

x:	2	3	4	5	6
$p(x)$:	0.1	k	0.15	0.2	0.3

This is a probability distribution

Find:
a) the value of k b) $P(X > 3)$ c) $P(X \leq 3)$ d) $P(X \geq 4.5)$
e) $P(2 < X < 5)$ f) $P(2 \leq X \leq 5)$ g) $F(4)$ h) $F(2.7)$

Solution

a) Since $\Sigma p(x) = 1$, then: $0.1 + k + 0.15 + 0.2 + 0.3 = 1 \Rightarrow k = 1 - 0.75 = 0.25$
b) $P(X > 3) = P(X = 4) + P(X = 5) + P(X = 6) = 0.15 + 0.2 + 0.3 = 0.65$
c) $P(X \leq 3) = P(X = 2) + P(X = 3) = 0.1 + 0.25 = 0.35$
d) $P(X \geq 4.5) = P(X = 5) + P(X = 6) = 0.2 + 0.3 = 0.5$
e) $P(2 < X < 5) = P(X = 3) + P(X = 4) = 0.25 + 0.15 = 0.4$
f) $P(2 \leq X \leq 5) = P(X = 2) + P(X = 3) + P(X = 4) + P(X = 5)$
$= 0.1 + 0.25 + 0.15 + 0.2 = 0.7$
g) $F(4) = P(X \leq 4) = 0.1 + 0.25 + 0.15 = 0.5$
h) $F(2.7) = P(X \leq 2.7) = P(X = 2) = 0.1$

2 Mean and Variance

You need to be able to use the following formulae to find the mean and variance of a discrete random variable:

Mean $= \mu = E(X) = \sum x.p(x)$ **(1)**

NB: $E(X)$ is the 'expectation of X'.

Variance $= \sigma^2 = E(X^2) - (E(X))^2$ **(2)**

Where $E(X^2) = \sum x^2.p(x)$ **(3)**

It is important that you realise that $E(X^2) \neq E(X) \times E(X)$. The only way to calculate $E(X^2)$ is by using formula **(3)**.

Example 2

Mobile telephone 'Pay & Go' vouchers are available in values of £5, £10, £20, £50 and £100 at the Mobile Shop. The random variable 'X = value of the mobile voucher sold in £s' has the following probability distribution:

x:	5	10	20	50	100
p(x):	0.25	0.35	0.25	0.12	0.03

Find:

a) $E(X)$ b) $Var(X)$ c) standard deviation of X

The Mobile Shop sells 2500 vouchers in a month.

d) Find the expected revenue from voucher sales in a month.

Solution

a) $E(X) = \sum x.p(x)$
$= (5 \times 0.25) + (10 \times 0.35) + (20 \times 0.25) + (50 \times 0.12) + (100 \times 0.03)$
$= 1.25 + 3.5 + 5 + 6 + 3 = £18.75$

b) $E(X^2) = \sum x^2.p(x)$
$= (5^2 \times 0.25) + (10^2 \times 0.35) + (20^2 \times 0.25) + (50^2 \times 0.12) + (100^2 \times 0.03)$
$= 6.25 + 35 + 100 + 300 + 300 = 741.25$

Hence $Var(X) = E(X^2) - (E(X))^2 = 741.25 - (18.75)^2 = 389.69\ldots$

c) Standard deviation $= \sigma_x = \sqrt{Var(X)} = \sqrt{389.68\ldots} = £19.74$ (4 sf)

d) On average, a voucher costs £18.75. So 2500 cost: $2500 \times 18.75 = £46875$

3 Linear Combinations of a Random Variable

You need to apply the following formulae to a linear combination of a random variable:

$E(aX \pm b) = aE(X) \pm b$

$Var(aX \pm b) = a^2\, Var(X)$

a, b are constants.

DAY 7

Example 3

You are given that $E(X) = -3$ and $E(X^2) = 15$. Find:

a) $E(3X)$ b) $E(4X + 3)$ c) $E(5 - 2X)$ d) $Var(X)$ e) $Var(3X)$
f) $Var(\frac{1}{2}X)$ g) $Var(5X - 9)$ h) $Var(3 - 2X)$

Solution

a) $E(3X) = 3E(X) = 3 \times -3 = -9$
b) $E(4X + 3) = 4E(X) + 3 = (4 \times -3) + 3 = -9$
c) $E(5 - 2X) = 5 - 2E(X) = 5 - (2 \times -3) = 5 + 6 = 11$
d) $Var(X) = E(X^2) - (E(X))^2 = 15 - (-3)^2 = 15 - 9 = 6$
e) $Var(3X) = 9Var(X) = 9 \times 6 = 54$
f) $Var(\frac{1}{2}X) = \frac{1}{4}Var(X) = \frac{1}{4} \times 6 = 1.5$
g) $Var(5X - 9) = 25Var(X) = 25 \times 6 = 150$
h) $Var(3 - 2X) = (-2)^2Var(X) = 4Var(X) = 4 \times 6 = 24$

4 Application of Discrete Random Variables

You can apply discrete random variables to problems involving probability.

Example 4

An urn contains 2 red balls, 7 blue balls and 3 yellow balls. Two balls are selected at random *without replacement* from the urn. The random variable X is defined as the number of yellow balls that are selected. Find the probability distribution for X.

Solution

- *Without replacement* means you keep the ball, once it has been taken from the urn.
- The number of yellow balls selected is either 0 or 1 or 2 $\Rightarrow x = 0, 1, 2$.
- When you select a ball at random it is either *yellow, y* or *not yellow y′*.
- There are 3 *yellow (y),* balls, and 9 *not yellow (y′)* balls in the urn.
- $P(X = 0) = P(y', y') = \frac{9}{12} \times \frac{8}{11} = \frac{6}{11}$
- $P(X = 1) = P(y, y') + P(y', y) = \left(\frac{9}{12} \times \frac{3}{11}\right) + \left(\frac{3}{12} \times \frac{9}{11}\right) = \frac{9}{22}$
- $P(X = 2) = P(y, y) = \frac{3}{12} \times \frac{2}{11} = \frac{1}{22}$
- Hence probability distribution is:

x:	0	1	2
$p(x)$:	$\frac{6}{11}$	$\frac{9}{22}$	$\frac{1}{22}$

Remember: these probabilities should add up to make 1

DAY 7

Discrete Random Variables

Have you improved?

1 A random variable X is defined by the probability function:

$$f(x) = \begin{cases} kx^2 & x = 1, 2, 3 \\ (8-x)k & x = 4, 5 \end{cases}$$

a) Find the value of k
b) Calculate $E(X)$
c) Find $Var(X)$

2 The random variable X has the following probability distribution:

x:	3	4	5	6	7	8
p(x):	0.15	a	2b	0.1	0.15	b

Given that $E(X) = 5.1$, find:
a) the values of the constants a and b
b) $Var(2X - 3)$.

3 A game is played where two fair 6-sided dice are rolled. The random variable X is defined as the smaller of the two scores on the dice.
a) Find the probability distribution of X.

The *profit*, Y (in £), when you play this game is equal to $2X - 7$.
b) Calculate (to the nearest penny) the expected profit when you play the game.
c) Comment on your answer.

1) First, find the probability distribution

1a) Use $\Sigma p(x) = 1$

1b) Use $E(X) = \Sigma x.p(x)$

1c) Find $E(X^2)$, then $Var(X)$

2a) Use $E(X)$ and $\Sigma p(x)$ formulae. Solve them!

2b) Find $Var(X)$ first

3a) Draw a grid to represent the possibilities

3b) Work out $E(X)$ from (a), then find $E(Y)$

3c) Look at the sign

Pure

1 a) Find the solutions to the equation $x^2 - 4x + 1 = 0$, giving your answers in the form $a + \sqrt{b}$, where a and b are integers.
b) Find the exact solution to the inequality $x^2 + 1 > 4x$.

2 a) Show that $(x + 2)$ is a factor of $2x^3 - 5x^2 - 14x + 8$
b) Hence find all the solutions of the equation $2x^3 - 5x^2 - 14x + 8 = 0$
c) Find all the solutions of the equation $2(2^{3x}) - 5(2^{2x}) - 14(2^x) + 8 = 0$

3 Solve the following equations in the interval: $-180° \leq x \leq 180°$:

a) $\sin(2x - 30) = -\frac{\sqrt{3}}{2}$

b) $2\cos^2 x = 3\sin x$

Mechanics

1 Particles A and B, each of mass 2 kg, are connected by a light inextensible string. Particle A is held at rest on a rough wedge with the string passing over a smooth pulley and particle B hanging freely, as shown.

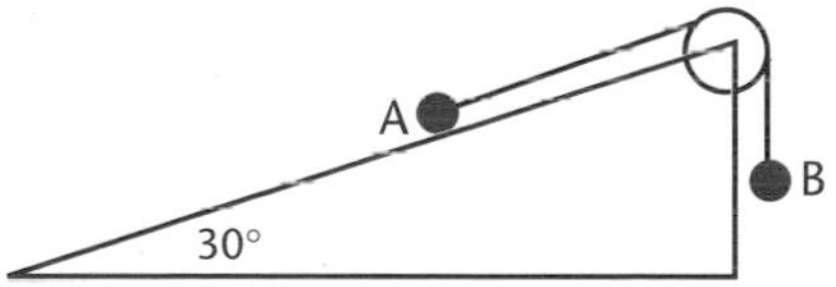

When A is released from rest, the particles move with acceleration $0.5\,\text{ms}^{-2}$. Find:
a) the tension in the string
b) the coefficient of friction between particle A and the plane.

2 Annette is standing at the point with position vector $3\mathbf{i} + 4\mathbf{j}$ when she hits a hockey ball with velocity $\mathbf{v}$.
Darsha, who was standing at the point with position vector $12\mathbf{i} + 2\mathbf{j}$, starts running with velocity $-2\mathbf{i} + \mathbf{j}$ when Annette hits the ball, and intercepts the ball 2 seconds later. Find the speed with which Annette hit the ball.

3 Particle A, of mass $3m$, is travelling with constant speed $2u$. It collides with particle B, of mass m, which is travelling with constant speed u in the same direction. After the collision, the particles move with speeds v and $3v$ respectively.

a) Find v, in terms of u.

b) Find the impulse exerted, in terms of m and u, by A on B in the collision.

Statistics

1 The following data show the weights (to the nearest pound) of some new-born babies.

Weight (pounds)	Number of babies
3–4	2
5–6	6
7	8
8	7
9–10	5

a) Find estimates for the mean and standard deviation of the weights, and explain why these are only estimates.

b) Use linear interpolation to find estimates for the median and interquartile range of the weights.

c) Explain the circumstances in which it may be preferable to use the median and interquartile range as measures of location and spread, rather than the mean and standard deviation.

d) Suggest a suitable diagram to represent the data, and give one advantage and one disadvantage of your suggested type of diagram.

2 The table below shows the number of students studying physics and/or mathematics in the Lower Sixth at Handways School.

	Studies physics	Does not study physics
Studies mathematics	21	27
Does not study mathematics	3	39

A student from the Lower Sixth at Handways School is selected at random.
The events M and P are defined as 'The selected student studies mathematics' and 'The selected student studies physics' respectively.

a) Explain what is meant by the events M' and $M|P'$.

b) Find the probabilities of the two events defined in part a) and state with reasons whether or not the events M and P are independent.

62.5% of those students who study mathematics are girls, and 50% of those who do not study mathematics are girls.

c) Show that the probability a randomly selected student is a girl is $\frac{51}{90}$.

Three students are selected at random.

d) Find the probability that one is a boy who studies mathematics, one is a girl who studies mathematics, and the third does not study mathematics.

Answers on page 112

Have you improved: Answers

Algebra

1a) Using formula: $x = \frac{6 \pm \sqrt{(36-8)}}{2} = \frac{6 \pm \sqrt{28}}{2} = \frac{6 \pm 2\sqrt{7}}{2} = 3 \pm \sqrt{7}$

b) $\underline{\quad 3-\sqrt{7} \qquad 3+\sqrt{7} \quad}$
Put in $x = 0$: $(0)^2 - 6(0) + 2 = 2$ +ve
$x = 3$: $(3)^2 - 6(3) + 2 = -7$ −ve
$x = 6$: $(6)^2 - 6(6) + 2 = 2$ +ve
So $\underline{\text{+ve} \quad 3-\sqrt{7} \quad \text{−ve} \quad 3+\sqrt{7} \quad \text{+ve}}$. So require $3-\sqrt{7} \le x \le 3+\sqrt{7}$

2 Let length = x. Then width = $x - 4$.
Perimeter = $2x + 2(x-4) = 4x - 8 \le 36$ **(1)**
Area = $x(x-4) \ge 60$ **(2)**
(1) $\Rightarrow 4x \le 44 \Rightarrow x \le 11$
(2) $\Rightarrow x^2 - 4x - 60 \ge 0 \Rightarrow (x-10)(x+6) \ge 0$, $\underline{\text{+ve} \quad -6 \quad \text{−ve} \quad 10 \quad \text{+ve}}$
So require $x \ge 10$ (since can't have negative length)
So combining **(1)** and **(2)**, require $x \ge 10$ and $x \le 11$, so $10 \le x \le 11$

3a) $-2x^2 - 20x + 15 \equiv A(x+B)^2 + C \equiv A(x^2 + 2Bx + B^2) + C \equiv Ax^2 + 2ABx + AB^2 + C$
Looking at x^2: $-2x^2 \equiv Ax^2 \quad \Rightarrow A = -2$
Looking at x: $-20x \equiv 2ABx \quad \Rightarrow B = 5$
Constants: $15 \equiv AB^2 + C \quad \Rightarrow 15 = -50 + C \Rightarrow C = 65$
So $-2x^2 - 20x + 15 \equiv -2(x+5)^2 - 65$

b) i) $-2(x+5)^2 + 65 + 31 = 0 \Rightarrow 2(x+5)^2 = 96 \Rightarrow (x+5)^2 = 48 \Rightarrow x + 5 = \pm\sqrt{48} = \Rightarrow x = -5 \pm 4\sqrt{3}$
ii) It is maximum when $(x+5)^2 = 0$, so maximum value is 65

4 $2^{x+1} = 2^x \times 2^1 = 2y$ $\qquad 2^{-x} = \frac{1}{2^x} = \frac{1}{y}$

So $2^{x+1} + 2^{-x} = 3 \Rightarrow 2y + \frac{1}{y} = 3 \Rightarrow 2y^2 + 1 = 3y \Rightarrow 2y^2 - 3y + 1 = 0 \Rightarrow (2y-1)(y-1) = 0 \Rightarrow y = \frac{1}{2}, 1$

But $y = 2^x$. $y = \frac{1}{2} \Rightarrow 2^x = \frac{1}{2} \Rightarrow x = -1$. $y = 1 \Rightarrow 2^x = 1 \Rightarrow x = 0$

5a) Try $x = 1$: $1^3 - 1^2 - 1 - 2 = -3$ so $(x-1)$ not a factor
Try $x = 2$: $2^3 - 2^2 - 2 - 2 = 0$ so $(x-2)$ is a factor
$x^3 - x^2 - x - 2 = (x-2)(x^2 + Ax + 1)$ Coeff. x^2: $-1 = -2 + A \Rightarrow A = 1$
So $x^3 - x^2 - x - 2 = (x-2)(x^2 + x + 1)$
But for $x^2 + x + 1$, $b^2 - 4ac = 1 - 4 < 0$, so no roots. So $x = 2$ is only root.

b) If $y = 2^x$, this reduces to $y^3 - y^2 - y - 2 = 0$. So only root is $y = 2 \Rightarrow 2^x = 2 \Rightarrow x = 1$

Coordinate Geometry

1a) gradient = $\frac{0-5}{0-2} = \frac{5}{2} \Rightarrow y = \frac{5}{2}x$

b) Equation WX $\Rightarrow y - 5 = -2(x-2) \Rightarrow y = -2x + 9$.
Grad WX = $-2 \Rightarrow$ Grad XY = $\frac{1}{2}$, because lines are perpendicular.

Equation XY $\Rightarrow y - 2 = \frac{1}{2}(x-6) \Rightarrow y = \frac{1}{2}x - 1$.

c) $-2x + 9 = \frac{1}{2}x - 1 \Rightarrow 10 = 2.5x \Rightarrow x = 4$. At $x = 4$, $y = -2(4) + 9 = 1$. So coordinates are X(4, 1)

d) Grad OW = Grad YZ = 5/2

Equation YZ $\Rightarrow$ $y-2=\frac{5}{2}(x-6)$ $\Rightarrow$ $y = \frac{5}{2}x-13$

When $y = 0$: $0 = \frac{5}{2}x-13$ $\Rightarrow$ $\frac{5}{2}x=13$ $\Rightarrow$ $x=\frac{26}{5}=5.2$

Hence length of base = 5.2 × 20 = 104 cm

2a) AB = $\sqrt{(-3-t)^2+(1--t)^2}=10$. Squaring gives: $(-3 - t)^2 + (1 + t)^2 = 100$

So $(-3 - t)(-3 - t) + (1 + t)(1+ t) = 100$ gives: $9 + 3t + 3t + t^2 + 1 + t + t + t^2 = 100$
Hence $2t^2 + 8t + 10 = 100 \Rightarrow 2t^2 + 8t - 90 = 0 \Rightarrow t^2 + 4t - 45 = 0 \Rightarrow (t + 9)(t - 5) = 0$
Giving $t = -9$ (reject $t > 0$) or $t = 5$ (accept) (So B(5, –5))

b) Gradient AB = $\frac{1--5}{-3-5}=-\frac{6}{8}=-\frac{3}{4}$. So equation AB $\Rightarrow y - 1 = -\frac{3}{4}(x--3)$

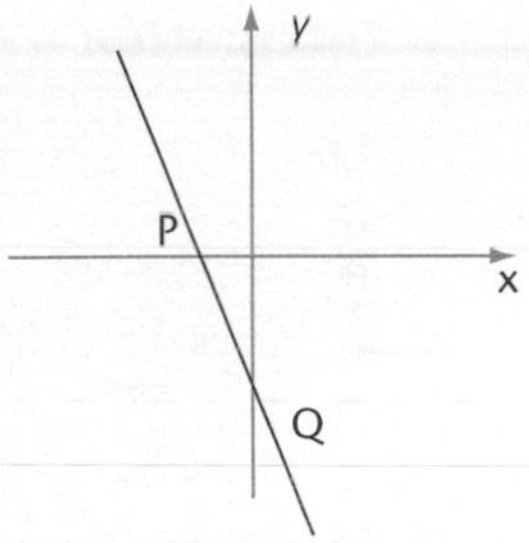

So $y = -\frac{3}{4}x-\frac{5}{4}$

c) P $\Rightarrow y = 0 \Rightarrow 0 = -\frac{3}{4}x-\frac{5}{4}$ $\Rightarrow x = -\frac{5}{3}$. So P(–5/3, 0)

Q $\Rightarrow x = 0 \Rightarrow y = -\frac{3}{4}(0)-\frac{5}{4}=-\frac{5}{4}$. So Q(0, $-\frac{5}{4}$)

Hence base = 5/3, height = 5/4 $\Rightarrow$ Area = $\frac{1}{2}(5/3)(5/4) = \frac{25}{24}$ (units)^2

Differentiation

1a) At $x = 2$, $y=\frac{4}{6}-2+\frac{1}{3}=-1$. So P(2, –1)

$\frac{dy}{dx}=\frac{1}{3}x-1$ $\Rightarrow$ When $x = 2$, $\frac{dy}{dx}=\frac{2}{3}-1=-\frac{1}{3}$ (= gradient of tangent)

$\Rightarrow$ Gradient of normal = $\frac{-1}{-1/3}=3$ $\Rightarrow$ equation of normal is $y - (-1) = 3(x - 2)$

So equation of normal is $y = 3x - 7$

b) Normal meets curve $\Rightarrow$ Normal = Curve $\Rightarrow$ $3x-7=\frac{x^2}{6}-x+\frac{1}{3}$ (× 6)

$18x - 42 = x^2 - 6x + 2 \Rightarrow x^2 - 24x + 44 = 0 \Rightarrow (x - 2)(x - 22) = 0$
So $x = 2$ (at P, known) or $x = 22$ (at Q) $\Rightarrow y = 3(22) - 7 = 59$. So Q(22, 59)

2a) At $x = 1$, $y = 4 - k$. So P(1, 4 – k). $\frac{dy}{dx}=4-2kx$. When $x = 1$, $\frac{dy}{dx}=4-2k$

Equation of tangent: $y - (4 - k) = (4 - 2k)(x - 1) \Rightarrow y - 4 + k = (4 - 2k)x - 4 + 2k$
So $y = (4 - 2k)x - 4 + 2k + 4 - k \Rightarrow y = (4 - 2k)x + k$ **(1)**
When $x = 0$, $y = 5 \Rightarrow 5 = k$

b) Substitute $k = 5$ into **(1)**: $y = (4 - 10)x + 5 \Rightarrow$ Eqn of tangent: $y = -6x + 5$

3a) Volume, $V = 2x^2h \Rightarrow 1728 = 2x^2h \Rightarrow \frac{1728}{2x^2}=h$ **(2)**

Surface area, $y = xh + xh + 2xh + 2xh + 2x^2 \Rightarrow y = 2x^2 + 6xh$ **(3)**

(2) into **(3)** gives: $y = 2x^2+6x\left(\frac{1728}{2x^2}\right)=2x^2+\left(\frac{6\times1728}{2x}\right)=2x^2+\frac{5184}{x}$

b) $\frac{dy}{dx} = 4x - \frac{5184}{x^2} = 0 \Rightarrow \frac{5184}{x^2} = 4x \Rightarrow x^3 = \frac{5184}{4} = 1296.$

$x = \sqrt[3]{1296} = 10.9027... = 10.9$ (*to* 3*sf*).

$\frac{d^2y}{dx^2} = 4 + \frac{10368}{x^3}$. At $x = 10.9027...$ $\frac{d^2y}{dx^2} = 4 + \frac{10368}{1296} = 12 \rangle 0.$

⇒ Minimum area at $x = 10.9$ cm.

c) $y_{min} = 2(10.9027..)^2 + \frac{5184}{10.9027..} = 713.216... = 713\text{cm}^2$ (to 3sf)

Integration

1a) $f(x) = \frac{(2x+1)(2x-1)}{\sqrt{x}} = \frac{4x^2-1}{x^{1/2}} = \frac{4x^2}{x^{1/2}} - \frac{1}{x^{1/2}} = 4x^{3/2} - x^{-1/2}$

$\Rightarrow A = 4\ \&\ B = -1$

b) $\int_1^2 4x^{3/2} - x^{-1/2}\, dx = \left[\frac{8}{5}x^{5/2} - 2x^{1/2}\right]_1^2 = \left(\frac{8}{5}(\sqrt{2})^5 - 2\sqrt{2}\right) - \left(\frac{8}{5} - 2\right)$

$= \frac{8}{5}(\sqrt{2}.\sqrt{2}.\sqrt{2}.\sqrt{2}.\sqrt{2}) - 2\sqrt{2} + \frac{2}{5} = \frac{32}{5}\sqrt{2} - 2\sqrt{2} + \frac{2}{5} = \frac{22}{5}\sqrt{2} + \frac{2}{5}$

$= \frac{2}{5}(11\sqrt{2}+1) \Rightarrow C = 11\ \&\ D = 1$

2a) $\frac{dy}{dx} = 2x - 3 \Rightarrow$ When $x = 3$, $\frac{dy}{dx} = 2(3) - 3 = 3$ (= gradient of tangent)

⇒ gradient of normal = $-\frac{1}{3}$ ⇒ Eqn of normal $y - 0 = -\frac{1}{3}(x-3) \Rightarrow y = -\frac{1}{3}x + 1$

b) Normal meets curve ⇒ normal = curve = $-\frac{1}{3}x + 1 = x^2 - 3x$ $(\times 3) \Rightarrow -x + 3 = 3x^2 - 9x$

$\Rightarrow 3x^2 - 8x - 3 = 0 \Rightarrow (x-3)(3x+1) = 0$. So $x = 3$ (at A, known) or $x = -\frac{1}{3}$ (at B)

$\Rightarrow y = -\frac{1}{3}(-\frac{1}{3}) + 1 = \frac{10}{9}$. So $B\left(-\frac{1}{3}, \frac{10}{9}\right)$

c) Let $C\left(-\frac{1}{3}, 0\right)$. So area ΔCBA = $\frac{1}{2}b.h$ $\therefore b = CA = 3 + 1/3 = 10/3$ and $h = 10/9$

So area ΔCBA = $\frac{1}{2}.\left(\frac{10}{3}\right).\left(\frac{10}{9}\right) = \frac{100}{54} = \frac{50}{27}$

Let area (S) = $\int_{-1/3}^{0} x^2 - 3x\, dx = \left[\frac{x^3}{3} - \frac{3x^2}{2}\right]_{-1/3}^{0} = (0) - \left(-\frac{1}{81} - \frac{1}{6}\right) = \frac{29}{162}$

Area shaded above x-axis = area ΔCBA – area (S) = $\frac{50}{27} - \frac{29}{162} = \frac{271}{162}$ **(1)**

$$\int_0^3 x^2 - 3x\,dx = \left[\frac{x^3}{3} - \frac{3x^2}{2}\right]_0^3 = \left(9 - \frac{27}{2}\right) - (0) = -\frac{9}{2}$$

Area shaded below x-axis = $-1 \times -\frac{9}{2} = \frac{9}{2}$ **(2)**

Area of region R = **(1)** + **(2)** = $\frac{271}{162} + \frac{9}{2} = 6\frac{14}{81} = 6.17$ (3 *sf*)

Trigonometry

1a)

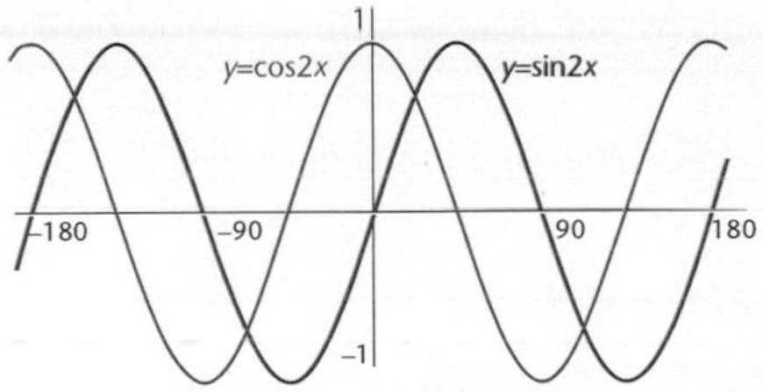

b) $\tan 2x = 1 \Rightarrow \frac{\sin 2x}{\cos 2x} = 1 \Rightarrow \sin 2x = \cos 2x$

This occurs when the graphs cross ⇒ 4 solutions

2 $2\sin x\cos x - \sin x - 2\cos x + 1 = (\sin x - 1)(2\cos x - 1) = 0$
So $\sin x = 1$ or $\cos x = 0.5$
So $x = 90°, 60°, 300°$

3 $\frac{\sin 2x}{\cos 2x} + 2\sin 2x = 0 \Rightarrow \sin 2x + 2\sin 2x\cos 2x = 0$

So $\sin 2x\,(1 + 2\cos 2x) = 0 \Rightarrow \sin 2x = 0$ or $\cos 2x = -0.5$
So $x = 0°, 90°, 180°, -90°, -180°, 60°, 120°, -60°, -120°$

4a) $\cos A = \sqrt{1 - \sin^2 A} = \sqrt{1 - \frac{1}{49}} = \sqrt{\frac{48}{49}} = \frac{4\sqrt{3}}{7}$

b) $\cos(90° - A) = \sin A = \frac{1}{7}$

c) $\sin(180° - A) = \sin A = \frac{1}{7}$

d) $\cos(180° + A) = -\cos A = -\frac{4\sqrt{3}}{7}$

5 See overleaf.

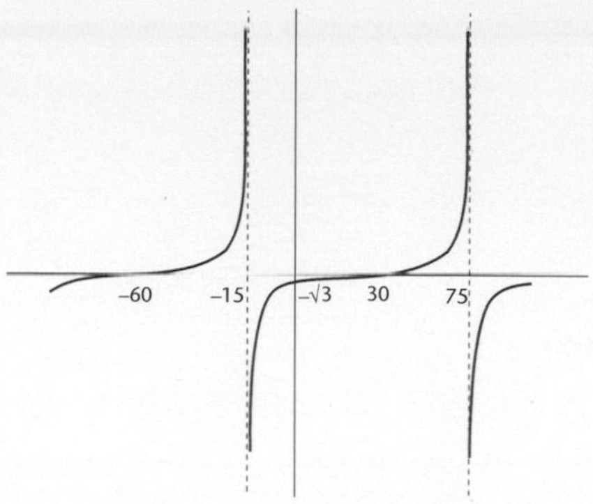

Series

1a) 7th term = 2(11th term) $\Rightarrow a + 6d = 2(a + 10d) \Rightarrow a + 6d = 2a + 20d$. So $a + 14d = 0$ **(1)**
5th term = 22.5 $\Rightarrow a + 4d = 22.5$ **(2)**
Solving simultaneously, i.e. **(1)** – **(2)**: $10d = -22.5$
So $d = -22.5/10 = -2.25$ **(1)** $\Rightarrow a = -14d = -14(-2.25) = 31.5$

b) $S_n = 189 \Rightarrow \frac{n}{2}(63 + (n-1)(-2.25)) = 189 \quad \Rightarrow n(63 - 2.25n + 2.25) = 378$

$\Rightarrow 63n - 2.25n^2 + 2.25n = 378 \Rightarrow 2.25n^2 - 65.25n + 378 = 0 \ (\div 2.25) \Rightarrow n^2 - 29n + 168 = 0$

$n = \frac{29 \pm \sqrt{841 - (4 \times 1 \times 168)}}{2} = \frac{29 \pm \sqrt{169}}{2} = \frac{29 \pm 13}{2} = 8 \text{ or } 21$

2a) Series: 7 + 14 + 21 + 28 + … + 742 + 749 $\Rightarrow$ AP: $a = 7$, $d = 7$, $n = 749/7 = 107$

Hence $S_{107} = \frac{107}{2}(14 + 106(7)) = 40446$

b) (Sum of non-multiples of 7) = (Sum of all numbers) – (Sum of multiples of seven)
So (Sum of all numbers) = 1 + 2 + 3 + 4 + … + 749 + 750 $\Rightarrow$ AP: $a = 1$, $d = 1$, $n = 750$

Hence $S_{750} = \frac{750}{2}(2 + 749(1)) = \frac{750 \times 751}{2} = 281625$

$\therefore$(Sum of non-multiples of 7) = 281625 – 40446 = 241179

3a) The £1200 that gets invested at beginning of each year, increases by 7% ≡ multiplied by 1.07
i) Value at end of yr 1 = 1200(1.07) = £1284 **(3)**
ii) In year 2: **(3)** remains invested and increases by a factor of 1.07. Then a further £1200 is invested at the beginning of the year, which in turn increases, by a factor of 1.07 by year-end.
$\therefore$ Value at end of yr 2 = $1200(1.07) + 1200(1.07)^2$
Hence, value at end yr 3 = $1200(1.07) + 1200(1.07)^2 + 1200(1.07)^3$ = £4127.93 (nearest pence)

b) $\therefore$ Total value = $1200(1.07) + 1200(1.07)^2 + 1200(1.07)^3 + \ldots + 1200(1.07)^9 + 1200(1.07)^{10}$

c) geometric

d) $a = 1200(1.07) = 1284$, $r = 1.07$, $n = 10$

$S_{10} = \frac{1284(1 - 1.07^{10})}{1 - 1.07} = 17740.319\ldots = £17740$ (*nearest pound*)

Equations of Motion

1a) accel = (gradient at t = 30) = $-\left(\frac{20-8}{60-0}\right) = -\frac{12}{60} = -0.2ms^{-2}$ $\Rightarrow$ deceleration = $0.2\,ms^{-2}$

b) Total distance = $\frac{1}{2}(20+8)60 + 8(T-60) + \frac{1}{2}(2T-T)8 = 840 + 8T - 480 + 4T$

$= 360 + 12T$. Now average speed = $\frac{\text{Total Distance}}{\text{Total Time}}$ $\Rightarrow 7.8 = \frac{360+12T}{2T}$

$\Rightarrow 15.6T = 360 + 12T \Rightarrow 3.6T = 360 \Rightarrow T = \frac{360}{3.6} = 100$ seconds

c) From sketch: Total time taken = $2T = 2 \times 100 = 200$ seconds

2a) Know: $u = 24.5$, $v = 0$, $a = -9.8$ Want: $s = h \Rightarrow v^2 = u^2 + 2as \Rightarrow 0 = 24.5^2 + 2(-9.8)h$

$\Rightarrow 19.6h = 24.5^2 \Rightarrow h = \frac{24.5^2}{19.6} = 30.625 \Rightarrow$ height above ground = 2 + 30.625

= 32.625 = 32.63 metres (2 dp)

b) Know: $u = 24.5$, $a = -9.8$, $s = 21.6 - 2 = 19.6$ Want: $t \Rightarrow s = ut + \frac{1}{2}at^2$

$\Rightarrow 19.6 = 24.5t + \frac{1}{2}(-9.8)t^2 \Rightarrow 19.6 = 24.5t - 4.9t^2 \Rightarrow 4.9t^2 - 24.5t + 19.6 = 0$ $(\div 4.9)$

$\Rightarrow t^2 - 5t + 4 = 0 \Rightarrow (t-4)(t-1) = 0 \Rightarrow t = 1$ or $t = 4$

$\Rightarrow$ Particle at least 21.6m above the ground when $1 \le t \le 4 \Rightarrow$ length of time $4 - 1 = 3$ s

c) Know: $s = -2$, $a = -9.8$, $u = 24.5$ Want: $t \Rightarrow s = ut + \frac{1}{2}at^2 \Rightarrow -2 = 24.5t + \frac{1}{2}(-9.8)t^2$

$\Rightarrow -2 = 24.5t - 4.9t^2 \Rightarrow 4.9t^2 - 24.5t - 2 = 0 \Rightarrow t = \frac{24.5 \pm \sqrt{24.5^2 - (4 \times 4.9 \times -2)}}{2 \times 4.9}$

$t = \frac{24.5 \pm \sqrt{639.45}}{9.8}$ $\Rightarrow t = -0.08$ s (reject!) or $t = 5.08$ s (accept!)

Equilibrium, Friction and Newton's Laws

1 $\theta = \tan^{-1}(3/4) \Rightarrow \sin\theta = 3/5$ and $\cos\theta = 4/5$

a) Resolving parallel and perpendicular to plane:
$2g \times 3/5 - F - P \times 4/5 = 0$
$2g \times 4/5 + P \times 3/5 - R = 0$
As equilibrium is limiting, $F = \mu R$.
So $6/5g - 4/5P = 0.1\,(8/5g + 3/5P)$
$P = 11.9$ N

b) Resolving as before and using $F = \mu R$:
$2g \times 3/5 + F - P \times 4/5 = 0$
$2g \times 4/5 + P \times 3/5 - R = 0$
$4/5P - 6/5g = 0.1(8/5g + 3/5P) \Rightarrow P = 18.0$ N

2 Resolving perpendicular and parallel to plane:
$R = mg\cos\theta$ **(1)**
$F = mg\sin\theta$ **(2)**

(2) ÷ (1): $\frac{F}{R} = \frac{mg\sin\theta}{mg\cos\theta} = \tan\theta$

But $F = \mu R$, so $\frac{F}{R} = \mu$ So $\mu = \tan\theta$.

3a) Resolving in direction of motion for each particle:
A: $T = 0.4a$ B: $0.6g - T = 0.6a$
Adding gives $0.6g = a = 5.88\,\text{ms}^{-2}$

b) Required force is resultant of the two forces shown:
From a), $T = 0.24g$.
Resultant = $\sqrt{2}T = 3.33\,\text{N}$ at 45° below horizontal

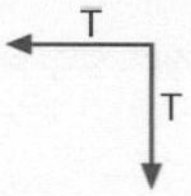

c) Using $2as = v^2 - u^2$:
$2 \times 0.6g \times 0.4 = v^2 \Rightarrow v = 2.17\,\text{ms}^{-1}$

Vectors in Mechanics

1a) For A: initial position vector is 8**j**, so $\mathbf{r}_A = 8\mathbf{j} + \mathbf{u}T$
For B: initial position vector is $-7\sqrt{2}\sin 45\mathbf{i} + 7\sqrt{2}\cos 45\mathbf{j} = -7\mathbf{i} + 7\mathbf{j}$
So $\mathbf{r}_B = -7\mathbf{i} + 7\mathbf{j} + \mathbf{v}T$

b) For A: $-17\mathbf{i} + 17\mathbf{j} = 8\mathbf{j} + 5\mathbf{u} \Rightarrow \mathbf{u} = -3.4\mathbf{i} + 1.8\mathbf{j}$ so speed is $\sqrt{(-3.4)^2 + 1.8^2} = 3.85\,\text{kmh}^{-1}$
For B: $-17\mathbf{i} + 17\mathbf{j} = -7\mathbf{i} + 7\mathbf{j} + 5\mathbf{v} \Rightarrow \mathbf{v} = -2\mathbf{i} + 2\mathbf{j}$
So speed is $\sqrt{(-2)^2 + 2^2} = \sqrt{8} = 2.83\,\text{kmh}^{-1}$

c) At 9am, $\mathbf{r}_A = 8\mathbf{j} + -3.4\mathbf{i} + 1.8\mathbf{j} = -3.4\mathbf{i} + 9.8\mathbf{j}$
$\mathbf{r}_B = -7\mathbf{i} + 7\mathbf{j} + -2\mathbf{i} + 2\mathbf{j} = -9\mathbf{i} + 9\mathbf{j}$
Distance between them = $\sqrt{(-3.4 - -9)^2 + (9.8 - 9)^2} = \sqrt{32} = 5.66\,\text{km}$

Time taken = distance ÷ speed = $\sqrt{32} \div \sqrt{8}$ = 2 hours ⇒ reaches A at 11am

2a) $\mathbf{r} = 4\mathbf{i} + 6\mathbf{j} + t(3\mathbf{i} - 2\mathbf{j}) = (4 + 3t)\mathbf{i} + (6 - 2t)\mathbf{j}$

b) $\frac{4+3t}{5} = \frac{6-2t}{1} \Rightarrow 4 + 3t = 30 - 10t \Rightarrow t = 2$

c) $\sqrt{(4+3t)^2 + (6-2t)^2} = \sqrt{13t^2 + 52}$

d) This is minimum when $t = 0$, so minimum distance is $\sqrt{52}$.

Momentum

1 $2m \times 3u = 3m \times v \Rightarrow v = 2u$

2a) $2 \times 6 = 3 \times v \Rightarrow v = 4\,\text{ms}^{-1}$

b) Look at 1 kg particle: change in momentum = $1 \times 4 - 0 = 4\,\text{kgms}^{-1}$

3a) $2m \times 12u + 3m \times -4u = 2m \times -2u + 3m \times v$
$12mu = -4mu + 3mv \Rightarrow v = \frac{16u}{3}$

b) Old kinetic energy = $\frac{1}{2} \times 2m \times (12u)^2 + \frac{1}{2} \times 3m \times (4u)^2 = 168mu^2$
New kinetic energy = $\frac{1}{2} \times 2m \times (2u)^2 + \frac{1}{2} \times 3m \times \left(\frac{16u}{3}\right)^2 = 46\frac{2}{3}mu^2$
Loss = old − new = $121\frac{1}{3}mu^2$

4 $3 \times 8 - 1 \times H = (3 + H) \times 1.5$
$24 - H = 4.5 + 1.5H$
$19.5 = 2.5H$
$7.8 = H$

5a) $6m \times 4 + 4m \times 3 = 6m \times v_A + 4m \times v_B \Rightarrow 36 = 6v_A + 4v_B$
Also, $v_B = v_A + 2$
Solving simultaneously gives $v_A = 2.8\,\text{ms}^{-1}$ and $v_B = 4.8\,\text{ms}^{-1}$
b) Impulse = new momentum – old momentum
$= 2.8 \times 6m - 4 \times 6m$
$= -7.2m\,\text{Ns}$

Descriptive Statistics

1a) True class boundaries are 39.5–49.5 etc., because weight is to nearest kg.
$\Sigma f = 100$, $\Sigma fx = 6042.5$, $\Sigma fx^2 = 368526.25$
mean = 6042.5 ÷ 100 = 60.425 kg, SD = $\sqrt{(368526.25/100 - 60.425^2)} = 5.84$ kg
These are only estimates because we haven't got the actual data values – the data have been grouped.

b) $\frac{1}{2}n = 50 \Rightarrow$ in 59.5–64.5 class. So median = $59.5 + \frac{50-42}{40} \times 5 = 60.5$

$\frac{1}{4}n = 25 \Rightarrow$ in 54.5–59.5 class. So LQ = $54.5 + \frac{25-12}{30} \times 5 = 56.6667$

$\frac{3}{4}n = 75 \Rightarrow$ in 59.5–64.5 class. So UQ = $59.5 + \frac{75-42}{40} \times 5 = 63.625$

So interquartile range = 6.9583

c) The median and mean are very similar; this reflects the fact that there are extreme values at each end of the range (the two people in the 39.5–49.5 category and the three people in the 69.5–89.5 category) so their effects on the mean cancel out. It also suggests the distribution is relatively symmetrical.

2a) Mean = $\frac{1+2+3+4+5}{5} = 3$, SD = $\sqrt{\frac{1^2+2^2+3^2+4^2+5^2}{5} - 3^2} = \sqrt{2} = 1.414\ldots$

b) i) We have added 10 to each value, so mean goes up by 10, so is now 13, and SD is unchanged at $\sqrt{2}$
ii) Have doubled each value, so mean is 6 and SD is $2\sqrt{2}$
iii) Have doubled each then added 30, so mean is 36 and SD is $2\sqrt{2}$, since not affected by adding 30
iv) Have multiplied by x and added y, so mean is $3x + y$, and SD is $x\sqrt{2}$

3a) Although data appears to be discrete, age is really a continuous variable so a histogram is suitable.
b) The overall distribution of ages can be seen easily.
c) This age group is actually $21 \le x < 26$, so of width 5 years. So 0.5 cm represents one year.
Frequency density is 3 ÷ 5 = 0.6. So height of column in cm is twice frequency density.
So for age 16–17, width is 1 cm, height is 2 cm; for age 20, width is 0.5 cm, height is 32 cm.

Have you improved: Answers

Probability

1a) P(positive) = 0.005 × 0.99 + 0.995 × 0.02 = 0.02485

b) This is P(disease|positive) = P(disease and positive)/P(positive) = 0.005 × 0.99 ÷ 0.02485 = 0.199

c) The probability of her having the disease for which she has tested positive is only about 20% – so most of the positive tests are wrong ones – so it is not a very good test!

2a) $A \cap B$ means A happens and B happens. $A \cap B'$ means A happens and B doesn't happen.
$P(A \cap B) + P(A \cap B')$ = prob[(A happens and B happens) or (A happens and B doesn't happen)] = P(A happens)

b) We know that all girls play netball or hockey or both. So girls either play just netball, netball and hockey, or just hockey. Just netball is $N \cap H'$. Just hockey is $N' \cap H$. Hockey and netball is $N \cap H$.
So $P(N \cap H') + P(N' \cap H) + P(N \cap H) = 1$

c) We know $P(N|H) = \frac{1}{4}$, $P(N) = \frac{11}{20}$. We want P(just hockey) = $P(N' \cap H)$

Try using $P(N \cap H') + P(N' \cap H) + P(N \cap H) = 1 \Rightarrow P(N' \cap H) = 1 - [P(N \cap H') + P(N \cap H)]$
But $P(N) = P(N \cap H) + P(N \cap H')$, from what was shown in part a)
So $P(N' \cap H) = 1 - P(N) = 9/20$

d) We know (from part a) that $P(H) = P(H \cap N) + P(H \cap N')$
$P(H \cap N') = \frac{9}{20}$, $P(H) = h$.

Have not yet used $P(N|H) = \dfrac{P(N \cap H)}{P(H)} \Rightarrow \frac{1}{4} = \dfrac{P(N \cap H)}{h} \Rightarrow P(N \cap H) = \frac{1}{4}h$

So, substituting in: $h = \frac{1}{4}h + \frac{9}{20} \Rightarrow \frac{3}{4}h = \frac{9}{20} \Rightarrow h = \frac{3}{5}$

e) P(not hockey|netball) = P(not hockey and netball) ÷ P(netball)
So we need $P(H' \cap N) = P(N) - P(H \cap N)$

But $P(N) = \frac{11}{20}$; $P(H \cap N) = \frac{1}{4}h = \frac{3}{20}$

So $P(H' \cap N) = \frac{11}{20} - \frac{3}{20} = \frac{2}{5}$

So P(not hockey| netball) = $\dfrac{\frac{2}{5}}{\frac{11}{20}} = \frac{8}{11}$

Correlation and Regression

1a) $\bar{x} = \dfrac{\sum x}{15} = 97.3 \Rightarrow \sum x = 1459.5$; $\bar{y} = \dfrac{\sum y}{15} = 138.2 \Rightarrow \sum y = 2073$

$$S_{xy} = 202009 - \frac{1459.5 \times 2073}{15} = 306.1,\ S_{xx} = 146587 - \frac{(1459.5)^2}{15} = 4577.65$$

$$S_{yy} = 289111 - \frac{(2073)^2}{15} = 2622.4 \Rightarrow r = \frac{306.1}{\sqrt{4577.65 \times 2622.4}} = 0.088\ (3dp)$$

b) Since the value of r is close to 0, we can say that from the sample, there is no linear correlation between a person's weight and their systolic blood pressure.

2a) See diagram

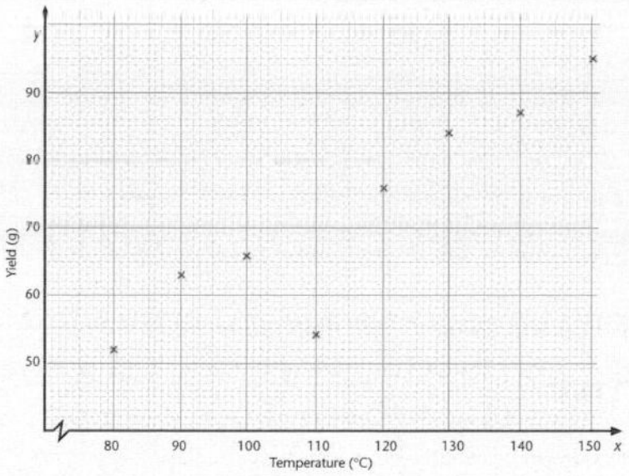

b) $x = 110$ °C, because this scatter point is further away from the predicted regression line.

c) From data; $\sum x^2 = 97900$, $\sum x = 810$, $n = 7$, $\sum y = 523$, $\sum xy = 62900$

$$S_{xy} = 62900 - \frac{810 \times 523}{7} = 2381.43..., \quad S_{xx} = 97900 - \frac{(810)^2}{7} = 4171.43...$$

$$b = \frac{2381.43}{4171.43} = 0.57089... \text{ and } a = \frac{523}{7} - (0.57089...)\frac{810}{7} = 8.654...$$

Hence $y = 8.65 + 0.571x$

d) Increase in mass of 15% $\Rightarrow$ Multiply old y by 1.15
i.e.: $y_{new} = 1.15y_{old}$ with $y_{old} = 8.65 + 0.571x \therefore y_{new} = 1.15(8.65 + 0.571x)$
So $y_{new} = 9.95 + 0.657x$

3a) α: Baby's weight at birth is 3.75 kg
β: Baby's weight gain per month is 0.28 kg

b) α: 230 units of gas are used then the temperature outside is 0 °C
β: Reduction of 7.5 units of gas used for every degree Celsius the temperature increases.

Discrete Random Variables

1a)

x:	1	2	3	4	5
$p(x)$:	k	$4k$	$9k$	$4k$	$3k$

$\sum p(x) = 1 \Rightarrow k + 4k + 9k + 4k + 3k = 1 \Rightarrow 21k = 1 \Rightarrow k = 1/21$

b) $E(X) = k + 8k + 27k + 16k + 15k = 67k = \frac{67}{21} = 3.19$ (3*sf*)

c) $E(X^2) = 1^2(k) + 2^2(4k) + 3^2(9k) + 4^2(4k) + 5^2(3k) = k + 16k + 81k + 64k + 75k = 237k$

$$= \frac{237}{21} = \frac{79}{7} \quad \Rightarrow \text{Var}(X) = \frac{79}{7} - \left(\frac{67}{21}\right)^2 = \frac{488}{441} = 1.11 \text{ (3 sf)}$$

2a) $\sum p(x) = 1 \Rightarrow 0.15 + a + 2b + 0.1 + 0.15 + b = 1 \Rightarrow a + 3b = 0.6$ **(1)**
$E(X) = 5.1 \Rightarrow 0.45 + 4a + 10b + 0.6 + 1.05 + 8b = 5.1 \Rightarrow 4a + 18b = 3$ **(2)**
Solving **(1)** and **(2)** simultaneously gives: $a = 0.3$ and $b = 0.1$

b) $E(X^2) = 9(0.15) + 16(0.3) + 25(0.2) + 36(0.1) + 49(0.15) + 64(0.1) = 28.5$
$\text{Var}(X) = 28.5 - (5.1)^2 = 2.49$. Hence $\text{Var}(2X - 3) = 4\text{Var}(X) = 4 \times 2.49 = 9.96$

3a)

X	**1**	**2**	**3**	**4**	**5**	**6**
1	1	1	1	1	1	1
2	1	2	2	2	2	2
3	1	2	3	3	3	3
4	1	2	3	4	4	4
5	1	2	3	4	5	5
6	1	2	3	4	5	6

Exam Practice: Answers

Hence probability distribution is:

x:	**1**	**2**	**3**	**4**	**5**	**6**
$p(x)$:	$\frac{11}{36}$	$\frac{9}{36}$	$\frac{7}{36}$	$\frac{5}{36}$	$\frac{3}{36}$	$\frac{1}{36}$

b) $E(X) = \frac{11}{36} + \frac{18}{36} + \frac{21}{36} + \frac{20}{36} + \frac{15}{36} + \frac{6}{36} = \frac{91}{36}$ $E(Y) = E(2X - 7) = 2E(X) - 7 = \left(2 \times \frac{91}{36}\right) - 7 = -\frac{35}{18} = -£1.94$

c) Over a long time, on average you make a loss of £1.94 per game.

Exam Practice

Pure

1a) $x = \frac{4 \pm \sqrt{16-4}}{2} = \frac{4 \pm \sqrt{12}}{2} = \frac{4 \pm 2\sqrt{3}}{2} = 2 \pm \sqrt{3}$

b) $x^2 - 4x + 1 > 0 \Rightarrow x > 2 + \sqrt{3}$ or $x < 2 - \sqrt{3}$

2a) Put in $x = -2$: $2(-2)^3 - 5(-2)^2 - 14(-2) + 8$
$= -16 - 20 + 28 + 8 = 0$

b) $2x^3 - 5x^2 - 14x + 8 \equiv (x + 2)(2x^2 + Ax + 4)$
$-5 = 4 + A \Rightarrow A = -9$
So $(x + 2)(2x^2 - 9x + 4) = 0$
$\Rightarrow (x + 2)(2x - 1)(x - 4) = 0$
So $x = -2, 0.5, 4$

c) $2^x = -2, 0.5, 4.$ So $x = -1, 2$

3a) $-390 \le 2x - 30 \le 330$
$2x - 30 = -60, -120, 300, 240$
$x = -15, -45, 165, 135$

b) $2 - 2\sin^2 x = 3\sin x \Rightarrow 2\sin^2 x + 3\sin x - 2 = 0$
$(2\sin x - 1)(\sin x + 2) = 0$
$\sin x = 0.5 \Rightarrow x = 30, 150$

Mechanics

1a) Considering particle B: $2g - T = 2 \times 0.5$
$T = 18.6$ N

b) Considering particle A:
$T - 2g\sin 30 - F = 2 \times 0.5$
$18.6 - 9.8 - F = 1 \Rightarrow F = 7.8$ N
Resolving perpendicular to plane:
$R = 2g \cos 30 = \sqrt{3}\, g$
$\mu = F \div R = 7.8 \div (\sqrt{3}g) = 0.460$

2 After 2 seconds, position vector of ball is:
$3\mathbf{i} + 4\mathbf{j} + 2\mathbf{v}$
Position vector of Darsha is:
$12\mathbf{i} + 2\mathbf{j} + 2(-2\mathbf{i} + \mathbf{j}) = 8\mathbf{i} + 4\mathbf{j}$
So $3\mathbf{i} + 4\mathbf{j} + 2\mathbf{v} = 8\mathbf{i} + 4\mathbf{j}$.
So $\mathbf{v} = 2.5\mathbf{i}$
So speed is 2.5 ms^{-1}

3a) $3m(2u) + mu = 3mv + m(3v)$
$7mu = 6mv \Rightarrow v = \frac{7}{6}u$

b) $m(3v) - mu = 3.5mu - mu = 2.5mu$

Statistics

1a) $\Sigma f = 28$, $\Sigma fx = 199.5$, $\Sigma fx^2 = 1497.25$
mean = 7.125, sd = 1.6455
They are only estimates because the actual data values are not known.

b) Cumulative frequencies 2, 8, 16, 23, 28
Median is in 14th position so
median $= 6.5 + \frac{14-8}{8} \times 1 = 7.25$ pounds

LQ is in 7th position so
LQ $= 4.5 + \frac{7-2}{6} \times 2 = 6\frac{1}{6}$ pounds

UQ is in 21st position so
UQ $= 7.5 + \frac{21-16}{7} \times 1 = 8\frac{3}{14}$ pounds

So interquartile range $= 2\frac{1}{21}$ pounds

c) If there are a small number of extreme values which will distort the mean and SD.

d) Histogram:
adv: can see shape of data distribution clearly
disadv: actual data values lost
Box plot: adv: can see spread of data/position of central 50% clearly
disadv: actual data values lost/lose detail of distribution

2a) M′: The selected student does not study Mathematics
M|P′: The selected student studies mathematics, given that they do not study physics

b) $p(M') = \frac{42}{90} = \frac{7}{15}$, $p(M|P') = \frac{27}{66}$
The events are not independent,
since $p(M) = \frac{8}{15} \ne p(M|P')$
(or using $p(M \text{ and } P) \ne p(M) \times p(P)$)

c) $0.625 \times 48 = 30$ girls study mathematics
$0.5 \times 42 = 21$ girls do not study mathematics
So altogether there are 51 girls $\Rightarrow$ probability is $\frac{51}{90}$

d) 18 boys study mathematics
So, if they are selected in given order, probability $= \frac{18}{90} \times \frac{30}{89} \times \frac{42}{88} = \frac{63}{1958}$
But they can be in 3! = 6 different orders, so probability $= \frac{189}{979}$ (= 0.1931)